EATER'S DIGEST

MICHAEL F. JACOBSON was born and raised in Chicago and graduated from the University of Chicago in 1965 with an AB in chemistry. Further studies in biology led him to the University of California at San Diego and finally to M.I.T., where he received his doctorate in microbiology in 1969.

Dr. Jacobson has served as a research associate of the Salk Institute and as a consultant to the Center for the Study of Responsive Law in Washington, D.C. In 1970 he joined several colleagues in the formation of the Center for Science in the Public Interest, where he is presently a co-director. Dr. Jacobson divides his time between investigating food additives and encouraging scientists to use their skills to benefit the public at large rather than vested interests. Other ongoing concerns of the Center are such consumer and environmental issues as strip mining, transportation, air and water pollution, and the safety of household chemicals.

Tax-deductible contributions to support these projects may be sent to the Center for Science in the Public Interest at 1779 Church Street, N.W., Washington, D.C. 20036.

EATER'S DIGEST

The Consumer's Factbook of Food Additives

MICHAEL F. JACOBSON

Foreword by Jean Mayer

ANCHOR BOOKS

Doubleday & Company, Inc.

Garden City, New York

Eater's Digest *was published
simultaneously in a hardcover edition
by Doubleday & Company, Inc.*

Anchor Books Edition: 1972

To the memory of
Colleen Meier
(1946–1972)

ACKNOWLEDGMENTS

It is a pleasure to acknowledge the encouragement and assistance I received from many persons while writing this book. Ralph Nader and James Turner both provided valuable initial encouragement. Marcy Benstock, John Esposito, Andrea Hricko, Debbie Luxenburg, Cindy Metzler, Jim Michael, Kathy Montague, Connie Jo Smith, Harrison Wellford, and other friends at the Center for Study of Responsive Law—in whose offices I began work on this book—supplied a steady stream of support, solace, companionship, and advice. Connie also helped me with the typing. Pamela Hackes and Leslie Conklin, then seniors at the National Cathedral School, helped gather information on artificial colorings and thickening agents. Nancy Stocker investigated problems connected with carrageenan.

The Center for Science in the Public Interest provided a stimulating and friendly atmosphere in which to complete this book. My fellow directors of the Center, Drs. Al Fritsch and Jim Sullivan, contributed much constructive criticism, helpful leads and, most importantly, a strong dose of infectious enthusiasm.

I should like to single out Dr. Samuel Epstein for his helpful and perceptive criticism of early drafts of this book. His standard of "impeccability" represented a lofty goal toward which I have striven but probably failed to attain. Sue Wellford, Peter Montague (Southwest Research and Information Center), Frensch Niegermeier, and Dr. Charles Kensler and his colleagues at Arthur D. Little also offered many useful comments after reading all or portions of various drafts. Naturally, though, the author himself must take full credit for all errors of fact or judgment that he has made in this book.

I am also indebted to Sibyl Weil, who did the drawings at the beginning of each chapter, to Barbara De Groot and Maureen Mahon of Doubleday, for their incisive criticism, valuable suggestions, and adept use of colored pencils, and to Marie Rodell, for her help in turning my manuscript into this book.

Finally, I should like to express my gratitude to The Salk Institute, San Diego, California, which, through its Council for Biology in Human Affairs, provided financial support during the period of research and writing of this book. I am especially grateful to Cy Levinthal (Columbia University), Harry Boardman, and Stuart Ross.

Contents

Chapter 3   A Close-up Look at Foods   197

FOREWORD

by Jean Mayer

Until very recently—as recently as 1969—the subject of food additives was considered by most Americans to be a pedestrian, dull, technological affair, of interest only to a few specialists working in that most unglamorous section of American business—the processed food industry. Suddenly, in a few short years, it has become one of the most controversial of all issues. Many Americans see the extensive use of food additives as raising not only profound health and economic implications, but also general questions about the very nature of our civilization. The combination of the heightened interest in nutrition and the U.S. food supply arising out of the First White House Conference on Food, Nutrition and Health, the shock waves generated by the ban on cyclamates and by subsequent, highly publicized FDA actions on food coloring and saccharin, the rise of the organic and nature food cults, the explosive growth of the consumer movements, have fed the growing passion with which the use of additives has been damned, or alternately praised. At times, the opposing factions have become so shrill as to sound like a quarrel of paranoiacs. Enemies of additives give us a nightmarish vision of a food industry dominated by a few malevolent men of great wealth who in their craze for profit are indiscriminately spraying their products with poisonous chemicals. By contrast, the Bull

Dr. Jean Mayer is Professor of Nutrition at Harvard University and served as Chairman of the First White House Conference on Food, Nutrition and Health.

Mooses of industry have perceived any criticism of additives as an attack on Science, Technology, the American Way of Life, and Western Civilization.

Intelligent citizens interested not in slogans but in understanding the complex problems involved in the use of additives will find the reading of Michael Jacobson's book highly rewarding. They will be exposed to the basic concepts underlying the establishment of a national food market that nourishes 230 million Americans and Canadians with domestic produce traveling thousands of miles in the long journey from the fields to the dining table and surviving time as well as distance, providing a much greater variety of food now in the dead of winter than could be obtained a century ago at the height of the growing season. And at the same time a drastic reduction has been achieved in the incidence of food-borne microbial disease, only recently one of the most widespread of human afflictions. The overflowing cornucopia is the result of technological advances in many fields, from agriculture to zoo technics. Food technology has been one of the most rapidly moving fronts, ever since François Appert developed industrial canning over a century ago. Preservatives have played an important role in the advance.

The women's liberation movement became possible when labor-saving devices freed adult females from many of the drudgeries of housekeeping. Refrigerators eliminated the need for daily food shopping, modern stoves and dishwashers reduced somewhat the time associated with the preparation of meals. The development of convenience foods, however, was the major quantum jump in freeing the housewife from the need of spending hours every day being the family cook. And many of the new foods, besides being nutritionally useful, tasted better than those produced by any but talented cooks. Again, food additives have played an indispensable role in the development of these time-savers. Without them, their texture, color, taste, smell, shelf-life and their very existence would be impossible.

These are obvious benefits. But there are risks and costs as well. Many new foods are excessively high in fat and salt, or in sugar, sources of "empty calories" made appetizing through the cosmetic effect of coloring matters and artificial flavors. Some of these "foods," further adorned with deliberately misleading names, beguile the consumer into thinking that he or she is consuming a nutritious food (or beverage—many of the "imitation fruit juices" are notorious examples) when essentially no protein, vitamin, or mineral is being ingested.

And, of course, there is always a risk in introducing one more unfamiliar chemical into our system—and many such chemicals have not been as thoroughly tested as they could and should have been. Not that new food additives are necessarily more dangerous than the thousands of normal constituents of established foods; cancer and other degenerative diseases existed ever since the human race started, millions of years before the first food additive was synthesized. But any added chemical of necessity increases the risk, albeit by a small amount.

Thus is born the concept of risk-benefit ratio. Is the risk that remains after adequate testing balanced by the advantages of a given additive in terms of protection against microbial or chemical spoilage, economics, or attractiveness? The consideration of risk-benefit ratio is basic to any intelligent discussion of any problem involving technology and society and is all too often ignored in the utterances of consumer advocates and industry spokesmen. Consideration of risk-benefit ratio makes it very clear that certain additives, or certain uses of certain additives, should be banned forthwith. In other cases, the additive presents, at the cost of a small risk, obvious "redeeming social significance" and should be retained, at least until a safer, more effective substitute is found.

Dr. Jacobson states the options in clear, dispassionate language and provides guidelines for the exploration of the dense crop of additives, singling out the most widespread, the most useful, the most questionable. As the review of additives hitherto "*G*enerally *R*ecognized *A*s *S*afe" (the "GRAS" list) demanded of FDA by the White House Conference continues, additional "surprising" findings are bound to occur. The reader of *Eater's Digest* will be prepared to understand the news and sometimes to anticipate it. More importantly, he will be able to maintain a balanced view of this difficult field. He will have an informed basis on which to decide when regulatory agencies need to be prodded to action, constructively criticized, or intelligently supported. Through its readers, this book will make an important contribution to the competence and the sanity of the public on which, in the long run, the main decision will rest.

EATER'S DIGEST

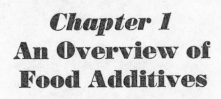

Chapter 1
An Overview of
Food Additives

INTRODUCTION

> "You do the dreaming—We'll do the rest.
> Imagination shouldn't be limited by today's
> technology. Go ahead! Dream up tomorrow's
> fantastic food ideas. We're already figuring
> out how you can do it, profitably."
>
> From *Food Product Development*, February/March 1971
> Advertisement, Durkee Industrial Foods Group

The lists of ingredients on food labels read like the index
of a chemistry textbook. A popular powdered "citrus"
drink contains:

> sugar, citric acid, gum arabic, natural and artificial orange
> flavors, cellulose gum, calcium phosphate, sodium citrate,
> ascorbic acid, hydrogenated vegetable oil, vitamin A, arti-
> ficial colors and butylated hydroxytoluene.

A nondairy whipped topping is concocted from:

> water, hydrogenated vegetable oil, sugar, vanilla, sodium
> caseinate, dextrose, polysorbate 60, sorbitan monostearate,
> carrageenan, guar gum and artificial coloring and fla-
> voring.

A popular cake mix contains a scientifically balanced
blend of:

> bleached cake flour, sugar, dextrose, corn syrup solids,
> shortening (with antioxidant), leavening, nonfat dry milk,
> propylene glycol monoesters, mono- and diglycerides,
> wheat starch, natural orange flavoring, dehydrated orange
> peel, glycerol, artificial vanilla flavoring, guar gum, citric
> acid and artificial coloring.

These manufactured, emulsified, stabilized, and preserved foods that line supermarket shelves are a far cry
from the fresh meat and vegetables that can still be found
around the fringes of the store. Today's foods contain
more than $485,000,000 worth of several thousand food
additives. In 1970 each of us consumed approximately
five pounds of these chemicals.[1]

"Just what are all these chemicals?", "Why are they
used?", "Are they safe?" are questions that everyone wonders about when they read food labels. The labels themselves are usually not much help in supplying us with
answers. "BHT added" says nothing about the chemical's
safety nor that it may not be necessary in the food. Are
the "vegetable gum stabilizers" sticky substances used to
prevent the contents of a package from rattling about?
What good is it to persons who don't have Ph.D.'s in
toxicology to be informed by a label that a food contains
sodium nitrite without also being told that nitrite is suspected of causing cancer and may actually serve no useful
purpose in that food?

For answers to our questions about food additives,
we have had three basically unappetizing choices. We
could read the popular books and articles, usually entitled something like "Poisons in Your Pantry," that paint
a uniformly dismal, and often misleading, picture of our
adulterated food supply and that condemn almost every
chemical created by God or man. Or we could write to
the manufacturer of a food or chemical and wait for the
inevitable reassuring response that runs something like
this: "Consumers of our product are well protected; all

----

[1] Estimates by Richard L. Hughes of Arthur D. Little Company,
as reported in *Chemical and Engineering News*, August 23, 1971
("Food Processing—Search for Growth," by Robert M. Hadsell).
These estimates are on the low side because they exclude sugar,
salt, vegetable oil, and similar substances that are used as additives
but considered to be foods.

of the ingredients in our products have been authorized by the United States Food and Drug Administration." Or we could write directly to our Federal Protector in Washington, the Food and Drug Administration (FDA), and receive the usual "informative booklet" which assures us that:

> If FDA scientists are satisfied that the additive may be used safely, a regulation will be issued permitting its use . . . If the evidence submitted is not convincing as to safety, the additive will not be permitted . . . The law prohibits the use of any chemical which would result in consumer deception.[2]

Eater's Digest offers what is hoped will be a satisfactory alternative to the above choices. In writing this book, the author studied the history, function, regulation, and safety of scores of common additives. He reviewed the scientific literature, interviewed scientists, scrutinized government reports, and corresponded with numerous manufacturers. *Eater's Digest* is intended to provide the average eater with an accurate, unbiased, and readable report on food additives.

The focus of the investigation of each food additive can be summarized in one word: safety. If a chemical additive was tested and found to be safe, as many have, there has been no hesitancy to state that fact in this book. On the other hand, if reliable scientific studies indicated that a chemical may be harmful or if a chemical has not been adequately studied, there was no reluctance to air these facts and, at times, to notify the Food and Drug Administration and the news media. The hazardous and untested additives are obviously the ones that readers should steer clear of. While this book was being written, the author filed petitions or complaints with the govern-

[2] *Food Additives: facts for consumers*, FDA, Washington, D.C.

ment about possible hazards associated with violet and
orange artificial food colorings, sodium nitrite, and brominated vegetable oil. Caffeine, quinine, BHT, BHA, and
propyl gallate are other additives that may be hazardous
or that serve no useful function in many foods. Additional
comments, warnings, advice, and recommendations are
liberally sprinkled throughout this book.

Why Are Additives Used?

Cooks were adding chemicals to their dishes long before
General Foods and Coca-Cola set up their prolific chemistry labs. Brine and smoke preservatives, seaweed and
starch thickeners, herb and spice flavorings, plant extract
colorings, and flavor enhancers made from dried fish have
been used for ages. Just about every naturally occurring
substance within reach has been added to food at one
time or another in man's perpetual effort to delight his
taste buds.

Today, new food additives, created by the powerful
combination of imaginative chemistry and aggressive food
technology, have replaced Nature as the good fairy that
helps fulfill (and create) our increasingly sophisticated
food wishes. Do people desire strawberry-flavored,
creamy topping that can be frozen or refrigerated and
that will not melt or wilt when placed on a pie months
later? If so, it is thickening agents, artificial flavoring,
preservatives, natural and artificial colorings, and synthetic emulsifiers to the rescue! Do campers fancy orange
beverages but not the chore of lugging around fresh, spoil-
able oranges, weighty bottles of juice, or melted frozen
concentrate? Then bring forth the powdered, colored,
flavored, preserved, fortified, instantly dissolving fruit of
a chemistry laboratory. The food and chemical industries
have proven that most any "food" that one can describe
can be created in a laboratory. This is what Durkee In-

dustrial Foods Group means when it advertises that "You do the dreaming—We'll do the rest. Imagination shouldn't be limited by today's technology. Go ahead! Dream up tomorrow's fantastic food ideas. We're already figuring out how you can do it, profitably."

Food additives have made possible more convenient foods, entirely new kinds of foods, and, in some cases, more economical foods. Consider bread: do you recall the days when we discarded half loaves that had become stale or moldy? That was before emulsifiers were used to prevent staling and before preservatives were used to prevent the growth of mold. Adding such substances to dough greatly extends the shelf life of bread and saves consumers the time and expense of buying bread every day or discarding spoiled bread. (The *taste* of bread, of course, is another matter.)

Oil-containing foods may go rancid when oxygen in the air reacts with polyunsaturated oils. Why not save the expense of throwing out spoiled mayonnaise or short-ening by simply adding a dash of antioxidant? Antioxidants, as well as other kinds of preservatives, can also lower manufacturing costs by permitting a company to produce, store, and ship extra large batches of food at a time and to stock up on ingredients when their cost is low.

The intelligent use of additives can overcome short-comings or annoying properties of traditional foods in many more ways than extending shelf life. Does ice cream melt too fast or does it develop an undesirable texture when it melts and then is refrozen? If so, add a vegetable gum stabilizer to hold it together and prevent ice crystals from forming. Does salt cake up? Add a silicate to keep the grains from sticking together. Is the meat tough and tasteless? Try a dash of tenderizer and a sprinkling of monosodium glutamate. Do dried apricots turn brown in the box? Just gas them with sulfur dioxide. The chemical cabinet contains remedies for almost every food problem.

Although they are not always thought of as food additives, minerals and vitamins are often used to fortify common foods with nutrients that are poorly supplied in the average diet. The addition of iron to bread and iodine to salt have contributed greatly to the health of the nation. Valuable vitamins and minerals lost in the production of white flour and white rice are partially replaced in "enriched" products. Factory-made foods, such as "instant breakfast," can be made nutritious by adding an assortment of vitamins, minerals, and proteins.

Aside from modifying familiar foods, the vast array of food chemicals has enabled manufacturers to create entirely new kinds of food. For busy people—or lazy people —ready-mix cakes, powdered juice bases, frozen dinners, dehydrated potatoes, and similar products, which are created, cooked, dried, powdered, or frozen at a distant factory, are time-saving miracles. Most of these easy-to-prepare dishes would not be possible or palatable without the generous use of flavorings, thickening agents, colorings, emulsifiers, and other additives.

We have been discussing food additives in functional terms, in much the way a company would respond to a consumer's inquiry as to why chemicals X, Y, and Z are present in a certain food. The answers are simple, brief, and leave little room for argument. Chemical X improves the food's shelf life; chemical Y adds to the nutritional value of the food; and chemical Z makes a smoother texture. Without denying the value of understanding, on this technical level, the purposes additives serve, we must proceed from the laboratory to company headquarters and answer the question "Why are food additives used?" on a more basic, economic level.

It is unlikely that, behind the closed boardroom doors of this nation's food and chemical companies, the managers are grappling with the problem of how to solve once and for all America's problems of hunger and malnutrition. More likely they are calculating how they can increase sales and asking how food additives can help. Some food additives make foods safer, others make foods cheaper, others make foods more or less nutritious; all food additives help companies make money, and that, in a nutshell, is why additives are usually used.[3] At least in the short run, it is more profitable for a company to advocate adding vitamins (which it happens to produce) to low-grade food than to tackle the nutritional problem at its roots by helping malnourished people get better jobs and more money so they can buy more nutritious food.

One of the oldest ways that merchants have of enhancing profits is to substitute an inexpensive ingredient for an expensive one. This frequently, but not always, decreases the food's nutritive value. Makers of ice cream substitute air and thickening agents for cream, while chocolate bar manufacturers replace a portion of expensive cocoa butter with lecithin. Soft drink bottlers promote artificially sweetened drinks, because saccharin (and, before it was banned, cyclamate) is far cheaper than sugar on a per bottle basis, not because such drinks help many persons lose weight; profit, not effectiveness, is the impetus. Synthetic vanillin replaces expensive vanilla as a flavoring. Protein or starch binders replace a fraction of

[3] Adelle Davis, America's favorite nutrition guru, says, "What's sad is that we're almost completely at the mercy of unethical manufacturers who don't give a damn about anybody's health. They have one goal, and that is to make money. Aside from that they couldn't care less about anybody's health." From the article "What Adelle Davis and Others Say About Eating Right," by James Trager, *Family Circle*, February 1972, page 50.

expensive meat in frankfurters, chili con carne, and frozen or dehydrated factory-prepared dinners. Thus, in the April 1971 issue of *Food Product Development,* Custom Food Products, Inc. advertised that:

> You can make a fortune in convenience foods . . . Our specially formulated binders turn out tender, juicy meats every time . . . with "rēcon" you'll win on the grill, or in the oven, every time.

In the same issue The Nestle Company described its way of helping food manufacturers cut costs:

> Nestle Vee-Kreme makes cream old-fashioned. Wherever you apply fresh dairy cream, replace it with Nestle Vee-Kreme. You will discover that Nestle Vee-Kreme successfully replaces a whole series of other ingredients in your formula. This saves you time, trouble . . . and money!

The whole spectrum of food additives, including such "good guys" as vitamins and minerals, can be used to improve a firm's profit picture. Let us examine several typical categories of food additives, focusing, as a sales department might, on their effect on corporate earnings.

Nutrients: criticizing the fortification of foods with vitamins and minerals has become almost as dangerous as slamming motherhood. But fortification must be recognized for the sales gimmick that it usually is; it is the food industry's equivalent to Hollywood's sex symbols. Some foods, such as white bread and white rice, have been stripped of most of their natural nutrients; then amid much advertising and self-congratulation manufacturers add back a few of the nutrients. Some companies are adding nutrients to breakfast cereals and cupcakes and advertising that the foods are nutritious. Nutritious, yes; wholesome, no. Most of these foods, such as ITT-Continental Baking's "Hostess" snacks and General Mills'

Kaboom cereal, are little more than vitamin-coated candy. The several vitamins they add are nutritious but do not cancel out the detrimental effects of sugar. Moreover, manufacturers frequently charge exorbitant prices for fortified products. "Total," for instance, is simply vitamin-coated "Wheaties." For their generosity in adding ½¢ worth of vitamins to twelve ounces of cereal, they add 18¢ (45 percent) to the retail price. Obviously, many food manufacturers are listening to Hoffman-La Roche, the nation's major vitamin manufacturer, when it advertises that:

> Recent research indicates that more and more mothers everywhere are aware of the need for vitamin and mineral fortification. And are even willing to pay a little extra for it. . . We can show you why it pays to fortify.

Some Americans need nutrients because they are not eating enough food, others because they are eating nutritionally inferior food. This problem of malnutrition can and should be solved at a more fundamental level than simply coating with vitamins and minerals whatever it is that people are currently eating.

Flavor enhancers: the food industry spent $24 million in 1970 on chemicals that accentuate the natural taste of foods. These chemicals, monosodium glutamate, maltol and ethyl maltol, and disodium guanylate and inosinate, do bring out the flavor. At the same time they enable a company to reduce the amount of natural ingredient in its product. It has gotten to the point where meat dishes can be made without meat. A major producer of disodium guanylate and disodium inosinate tells potential customers how they can use "Ribotide" to increase their profits:

> Is your flavor thinking old-fashioned? If you are locked-in on flavor enhancement, "Ribotide" is the best way to break

out, improve flavor, avoid problems. And save substantial sums of money, too.

Thickening agents and stabilizers: some companies have the integrity and patience to follow a traditional recipe or develop a new one and come out with a food that has a satisfying texture and consistency. Others use inferior ingredients or poor manufacturing techniques and rely on food additives to avoid a watery, lumpy, or crystalline product. This is most obvious in ice cream and yogurt production: some manufacturers need a shopping list of additives, others not a single one—and the taste test usually tells them apart in an instant.

Antioxidants: these chemicals are added to oil-containing foods to prevent the oil from going rancid and spoiling the food. "Enter the ageless age" advertises one manufacturer of the synthetic antioxidant butylated hydroxyanisole (BHA). Rancid-proof foods enable a manufacturer to ship larger quantities at a time, cutting down on shipping costs. The merchant never loses his investment on a slow-moving item if the item never loses its apparent freshness (whether it loses its vitamins is another question).

Meat analogs: 1970–1979 is likely to become the decade of imitation meat. Chemical companies have devised ways of processing soy protein that give it the taste, appearance, texture, and nutritive value of meat. They are currently at the beef stew and meat loaf stage but are working feverishly toward the era of imitation steaks. In all probability any American who eats in cafeterias has eaten meat loaf or hamburger that was composed partially of soy protein.

The profit potential of imitation meat is described in blunt terms in trade magazine advertisements:

Cargill Soy Products add nutritious protein to your finished product—at a remarkably low cost. *Your* customers get this extra quality bonus at less cost to you.

Meat is beautiful. Delicious. High in protein. Scarce. And Expensive. TVP brand (textured protein product) looks like meat. Can be flavored like meat. It's high in protein like meat. But it's plentiful, and just a fraction of the cost of meat.

A manufacturer, hospital cafeteria, or restaurant can save 20 to 25 percent by substituting soy protein for 30 percent of its beef. Rarely will such savings be reflected in the retail price of the food.

The most effective way that a manufacturer has of converting food additives to money is to use them to develop a new product. Food additives enable a chemist to translate an advertising man's wish into an actual product. The vast variety of industry-developed, government-approved additives makes possible foods that are immune to germs, that retain their consistency for months or years, whose artificial coloring does not fade in storage, that contain oil that will never go rancid, or that have unusual textures or shapes.

One food industry man said:

The marketing of many standard food items—the bread-and-butter items that have been around for years—has by now degenerated into a profitless price squeeze. Therefore, food companies simply must make innovations to maintain sales growth.[4]

New products mean money because they don't have any competition for a lengthy period of time. In this period they can become established as *the* brand to buy. The relative lack of competition enables a manufacturer to

[4] Anonymous, quoted in "Food Additives," H. J. Sanders, *Chemical and Engineering News,* October 10, 1966, page 118.

set as high a price as the public lets him get away with. Check the price per ounce of such new items as party snacks, dehydrated or frozen meals, vitamin-enriched breakfast cereals, and canned toppings or puddings. Then check the label for the list of additives that made these products possible. The prices are usually extraordinarily high compared to traditional foods (even steak!) despite the low cost of the major ingredients (vegetable oil, water, sugar, vitamins, emulsifiers, thickeners).

Eventually, of course, competitors will arrive, but by that time the research department will probably have developed a new food that makes the last one obsolescent, and the cycle can begin anew. Edible obsolescence. Frozen vegetables were once innovative and profitable, but the market got crowded and profits declined. The big producers began pushing new angles: plastic boiling bags, "international" flavors, and so on. Because housewives are not out in the streets demanding these new products, sales departments must shift into gear with heavy advertising expenditures to create the demand. Thus, the price of new foods is inflated an extra couple of notches because of bloated advertising budgets. It is interesting to note that the breakfast cereal industry, which is constantly offering new cereals "to meet today's modern needs," devotes about 13 percent of its income to advertising.

In the never-ending search for the Great New Product food technologists inevitably reach deep into their bag of food additive tricks or invent new ones if necessary. If some of these chemicals have not been exhaustively tested and exonerated of every foreseeable hazard, well, that is just a trade-off the public will have to make if it wants the time-saving conveniences and tasty new products that the food and chemical industries can create. As described more fully in the coming pages, not all additives have been systematically tested. Many chemicals have not been studied for their ability to cause cancer. Even fewer have been tested on pregnant animals to see if they cause birth defects. Virtually none have been screened for genetic effects.

While we are on the subject of safety, we ought to examine briefly the two extreme positions. The authors of "Poisons in Your Pantry" type books all too often fall into the trap of slandering *every* chemical used in food. Sometimes the slander gets so crude that a chemical is indicted just because its manufacturer also produces explosives or pesticides; obviously, the safety of one product has no bearing on the safety of the other. More commonly, these writers condemn a chemical because its safety has not been proved. We must recognize that many useful chemicals can never be proved absolutely safe—scientists can gather good evidence that it is not harmful, but proof of safety is impossible. Other traps that instant experts fall into are citing ancient scientific reports that have since been disproved, or citing unreliable scientists, or taking accurate quotes out of context. No chemical can be proved totally safe under all conditions for all persons; the best we can get, the most we can demand, and the least we must settle for is solid evidence that a hazard does not exist.

On the opposite side of the fence, the $100 billion a year food industry fills our magazines, newspapers, and air waves with a constant stream of propaganda criticizing those who dare question the safety of our foods. Their standard argument, in a nutshell, runs:

People, natural foods, air, and water are made of chemicals.
All of these things are safe.
Food additives are chemicals.
Therefore food additives are safe.

The Manufacturing Chemists Association's pamphlet entitled "Good Morning! Your 'Breakfast Chemicals'" lists some of the chemicals in coffee, eggs, milk, and other foods, then editorializes cheerily:

> So, you see, not only your favorite foods—but your best friends too—are chemicals!

Hartley W. Howard, the technical director of Borden, Inc., expands on that theme with this advice:

> Consumers must learn that *chemical* is not a dirty word; that our very bodies consist of nothing but *chemicals*—maintained and replenished by the *chemicals* in the foods we eat, the *chemicals* in the water we drink, and the *chemicals* of the air we breathe.
>
> Once consumers learn that those substances with the unfamiliar names are but a small addition to the vast number of chemicals provided by the more familiar food ingredients we will have made a considerable step forward.[5]

To argue that all food additives have been, are, and will be safe is as ridiculous as arguing that they are all dangerous. The Federal Government has banned a whole flock of food additives in the past half century. Twenty-five chemicals that have been banished to the food-chemical cemetery are described briefly in Appendix 1. The mere mention of some of these chemicals—agene, cyclamate, safrole, NDGA, butter yellow—sends chills up the spines of those who are familiar with their toxicities. It is astonishing that with all these skeletons in their closets industry spokesmen repeat their shopworn homilies with such innocence and sincerity and reject so vehemently the possibility that chemicals currently in our food supply may be dangerous. Although people and air

[5] "What Useful Purpose Is Served by Quantitative Ingredient Labeling," *Food Product Development*, June–July 1971, page 34.

are made of chemicals, food additives can still be dangerous.

The author's general rule on the safety of food additives is that there is no general rule. Some are safe, others are hazardous. Some are safe at low concentrations, but toxic at high concentrations. Some are themselves safe but make possible the fabrication of unwholesome foods. Each chemical must be examined individually.

A few chemicals, preservatives in particular, may introduce into our foods a small hazard, but one that is worth tolerating because the chemicals also make important contributions to public health. For instance, one reason that manufacturers add sodium nitrite to canned ham and cold cuts is to prevent the growth of bacteria that cause botulism, a deadly variety of food poisoning. There is a slight risk, however, that nitrite can contribute to cancer. At present the FDA believes that the health insurance we receive from nitrite outweighs the slim chance that the additive is hazardous. The best remedy for situations in which beneficial additives introduce small hazards is not to ban all preservatives, but to minimize the use of hazardous ones and to develop new ones that are as effective as, but less hazardous than, those currently under suspicion.

Some additives introduce a hazard without offering offsetting benefits. The use of sodium nitrite in baby foods, for instance, has a slight effect on the color and taste but is of no value whatsoever to the baby. It does not act as a preservative, nor does it increase the nutritional value or palatability. Gerber and Beech-Nut use nitrite in some of their products. Since nitrite has been suspected of contributing to cancer, mothers and public officials are clamoring for it to be removed from baby foods, but it should not have been tolerated even before there was a question of safety. A substance has no business being in our food if it does not benefit the consumer; the issue is that simple.

A second additive that was added to baby food to satisfy the parents' taste buds rather than the baby's health was monosodium glutamate (MSG), the flavor enhancer. Manufacturers stopped using it only in the face of massive public pressure generated by scientists, congressmen, and Ralph Nader when they discovered that it might harm the infant. MSG is a good example of a substance that scientists always considered innocuous, but that was shown by new research to be hazardous to infants. It, too, should never have been in baby food.

A third example of unnecessary additives involves antioxidants. Synthetic chemicals, including BHA (butylated hydroxyanisole), BHT (butylated hydroxytoluene), and propyl gallate, are added to many brands of vegetable oil and shortening to prevent them from going rancid. Some companies, though, have discovered that improved manufacturing procedures eliminate the need for these additives, an indication that antioxidants are not necessary in oils and shortening and should not be employed by any manufacturer.

Artificial colorings can add nothing to food but hazard and a cosmetically improved appearance. Candies and breakfast cereals are routinely colored simply for the sake of variety. Not only do colorings add hazard without matching benefits, but sometimes consumers are not even told of their presence. Florida citrus growers treat oranges with a coal tar dye to cover up their natural splotchy green color. Sodium nitrite and sodium erythorbate contribute to the pink color of cured meats. Red potatoes and sweet potatoes are sometimes dyed red. In all these cases, in spite of federal labeling laws, the consumer is usually not told that the food has been artificially colored. As discussed in Chapter 2, two colorings may be carcinogenic; none has been adequately tested.

Laws currently on the books forbid manufacturers to use dangerous food additives or harmless additives that promote deceptions, but the government cannot ordinarily prevent a manufacturer from adding a completely superfluous substance to its products. The American public should not be exposed to the risks, however small, presented by unnecessary food additives. Chemicals that now may appear to be safe may in the future turn out to cause cancer, birth defects, or other tragedies. Clearly, Congress must give the Food and Drug Administration the power to prevent uses of food additives that do not benefit the consumer.

In addition to problems related to safety, food additives may be undesirable for nutritional reasons. Many of the new foods that have been made possible by additives contain only carbohydrate, fat, water, and additives—*no protein, no vitamins, no minerals*. While these products may look irresistible, taste delicious, and take just seconds to prepare, they are totally devoid of nutrients. Too many Americans have traded nutrition for convenience, superficial attractiveness, and, possibly, poor health. Margaret Mead, the noted anthropologist, has expressed concern about "foods guaranteed not to nourish you . . . We have built up in this country such peculiar attitudes about food and nutrition that we now have a large group of starving people in the U.S., which would have been inconceivable twenty years ago."[6]

Nowadays, all too many parents pack lunchboxes with Fritos instead of oranges; after school, children reach for Fiddle Faddles instead of an apple; sales of artificially

[6] *Advertising Age,* May 4, 1970.

colored, sweetened, thickened, emulsified nonnutritive products, such as soft drinks and synthetic pies, are soaring, while the consumption of milk and fresh fruits and vegetables has declined. Factory-packed frozen or dehydrated dinners are low in protein, vitamins, and minerals when compared to old-fashioned home-cooked meals. It is no wonder that the Department of Agriculture's dietary surveys showed that a smaller fraction of Americans received adequate amounts of nutrients in 1965 than in 1955.

Food additives can save consumers time, raise corporate profits, create new products and improve old ones, and make foods more nutritious. On the other hand, many additives have not been adequately tested, some persons may be sensitive to certain additives, many factory-made foods made with additives are devoid of nutrients, and some chemicals may be used solely for cosmetic, deceptive, or irrational purposes. *Eater's Digest* will provide you with detailed information to help you avoid dangerous additives and to enable you to stop worrying about safe additives. Recognizing, though, the difficulty in remembering the risks and benefits of each and every additive, let me end this section by suggesting a rule of thumb on eating. For the safest, most nutritious, and most economical meals, base your diet on fresh or frozen meat, poultry, and produce, low-fat dairy products, and whole grains; eat a varied diet, and take special care to avoid foods with artificial colorings and antioxidants.

The Scope of This Book

Several thousand different substances are intentionally added to our foods. These substances range from citric acid and mono- and diglycerides, millions of pounds of which are used annually, to flavorings, the usage of which amounts to tens or hundreds of pounds per year per substance. They range from the sugar that sweetens our cereal to the wax that coats our cucumbers. Because this book would never have been completed if each and every one of the additives were evaluated, only the major additives were investigated. The two main criteria for selecting additives to be included in the study were (a) importance and (b) whether a reader might want to know about the chemical. These considerations guaranteed the inclusion of all the common additives and almost all the additives mentioned on food labels or in Standards of Identity (see Chapter 3). Emulsifiers, thickening agents, colorings, preservatives, flour treatment agents, and acids are discussed. Excluded are all of the herbs and spices and most flavorings. Also not covered are the accidental, but extremely important, contaminants: pesticides, hormones, antibiotics, and processing aids.

Official tabulations of food additives include nutrients and natural substances such as vitamins, amino acids, iron, milk protein, cellulose, salt, and sugar. Most of these everyday chemicals, except those which have functions other than making food tastier or more nutritious, get short shrift or are not discussed at all (see Appendix 2).

Several hundred different species of plants or extracts of plants are used as spices and, under the law, are food additives. Most of these substances, such as cloves, vanilla extract, and thyme, have been in use for ages and are probably innocuous. Few herbs and spices, however, have been studied in the laboratory. Until scientists conduct tests, we cannot state with any assurance that these substances are safe. The suggestion that the spice cabinet may contain hazardous compounds is more than just an idle worry. Sassafras, the spice which was used to flavor root beer, was found to contain a cancer-causing chemical and in 1960 was banned from foods (except when the offending ingredient, safrole, is removed). In 1968 another flavoring, oil of calamus, was found to cause cancer and was banned from food. (See the discussion of "artificial flavoring" in Chapter 2.)

In these days of mechanization and huge corporate farms, a sizable fraction of our farm-grown food is contaminated with potentially dangerous chemicals. Grains are cultivated in heavily fertilized fields, which may also have been sprayed with weed-killers; after harvest the grain may be treated with insecticides or fungicides. Cattle and poultry may eat fodder that contains hormones or antibiotics to spur their growth. Packaging materials, be they burlap sacks, treated papers, or detergent-washed cans, may further contaminate the food.

Manufactured foods encounter many different objects and chemicals before they reach the dinner table and are little purer than farm produce. Fortunately, though, manufacturing contaminants are rarely as biologically active as pesticides and hormones. Small amounts of the materials and machinery used to produce, transport, or store partially prepared food (or food ingredients) may wind up in the final product. All of these detergents, solvents, lubricating oils, defoamers, adhesives, plasticizers, resins, rubber catalysts, textile fibers, plastics, and preservatives, of which there are thousands, are considered by the Food and Drug Administration to be food additives (see Code of Federal Regulations, Title 21, sections 121.2500–2613).

Pesticides, hormones, antibiotics, and processing materials may persist in active form even after food is cooked and digested, but, of course, they are never listed on labels as additives because they are never intended to be in the food. For more extensive discussions of these substances, the reader is referred to books listed in the bibliography.

How Are Food Additives Tested?

Most persons are concerned about food additives because they seriously doubt the safety of the chemicals that are listed on their food packages. Every few years the FDA, university scientists, or muckrakers produce evidence that a supposedly safe substance is really hazardous. Cyclamate, agene, dulcin, butter yellow, safrole: these and other once popular, now banished, food additives portend future scandals and worries. In Chapter 2 we evaluate the safety of the more common additives. The present section is a primer on the biological tests that are—or should be—conducted to determine whether a food additive is safe.

A manufacturer who plans to introduce a new food additive bears the responsibility of conducting scientific tests on animals to establish the safety of the chemical. The substance must be shown to be free from hazards before it may be used in foods. Logical and necessary as this procedure may seem, it is actually of remarkably recent origin, having been translated into law only in 1958 (in 1960 for color additives). Prior to 1958 the situation was reversed: A manufacturer could put anything he pleased into food; the Food and Drug Administration could compel him to remove a chemical only after it had proved that the chemical was dangerous. Not surprisingly, the understaffed, slow-moving agency had a difficult time keeping the food supply free of hazardous chemicals, particularly chemicals, such as carcinogens, whose effects cannot be seen until long after exposure.

The weakest part of the 1958 food additive law pertains to those chemicals that were already in our food when the law was passed. Such additives were accepted as being safe solely on the basis of their having been used in food for many years; scientific evidence of safety was not required. These chemicals have been termed "generally recognized as safe" or GRAS (pronounced grass) for short. Most GRAS chemicals can be used in any food at any concentration. Manufacturers were authorized by the law to add new chemicals to the GRAS list after 1958, but only if scientific tests showed that the chemicals were safe; the manufacturers, not the Food and Drug Administration, were allowed to decide what constitutes adequate scientific evidence. GRAS chemicals are controlled by the FDA only after they are in the food. If the FDA has reason to believe that a chemical may be dangerous, it can order it off the GRAS list and demand that scientific tests be conducted (as happened with brominated vegetable oil) or ban its use in food (as happened with cyclamate). More information about GRAS chemicals is in the Glossary, pages 244–45.

After a manufacturer has tested his proposed additive on animals, he sends a food additive petition to the FDA. The petitioner must state the composition of the additive, its function, the amounts to be used in food, in what foods it will be used, the testing procedures and the results of the tests. FDA scientists study the petition and, if the chemical appears safe, the FDA announces its approval of the additive and the conditions of its proper use. The public has sixty days to object to FDA's decision, after which time foodmakers can use the new additive. But intelligent comment on a petition is usually difficult, because neither the FDA nor the manufacturer is likely to allow a potential critic to examine the petition. Note that the FDA's role is limited to evaluating the manufacturer's data; the FDA does not perform its own tests to verify the information that the petitioner submitted.

As I discovered while doing research for this book, citizens cannot examine the data on which the safety of a food additive is based. A manufacturer tests a chemical and forwards the results, or at least the results of selected experiments, to the FDA. The FDA labels the petition "confidential" and does not permit the people who will consume the chemical—i.e. you and I—to study the safety data.[7] The best you can do is obtain a summary compiled by FDA staff members, usually after months of delay. Whether the summary is objective or emphasizes information that buttresses FDA's past decisions cannot be determined without having access to the complete petition. Anyone who asks the FDA for a summary of studies in their files should expect to wait two to twelve months for the data.

As we have become more aware of the range of harmful effects a chemical may have, the FDA has required manufacturers to conduct more stringent tests on food additives. Prior to the 1950s, most toxicological studies were of short duration and could detect only out-and-out poisons. The amount of chemical needed to kill an animal was frequently accepted as an index of hazard. In the last twenty years we have become increasingly concerned about subtle effects and long-range effects of the food additives and other chemicals to which we are exposed daily, and the sophistication of testing has improved markedly (but there is still plenty of room for improvement). Scientists test more animals for longer periods of time and look for a wider variety of effects.

The most difficult effects to detect are those that do not show up for months or years after a person is exposed to a chemical. A food additive that had no immediately toxic effects but that causes birth defects, cancer, or mutations might escape detection indefinitely, because there

[7] The FDA will open its records to a citizen if a manufacturer sends a letter of authorization; a dozen companies rejected my request for permission to look at their food additive petitions. Most advised me to talk to people at FDA!

is no simple way of associating the ingestion of the chemical in 1971 with the birth of a deformed child in 1972, the occurrence of liver cancer in 1992 or a case of hemophilia in 2052. Extensive testing on rats, mice, dogs, and other laboratory animals is the only practical means of weeding out food additives before they cause subtle or long-range damage in people.

The most important of the tests used to investigate food additives is the chronic, or lifetime, feeding study. This test is designed to detect carcinogens (chemicals that cause cancer) and chemicals that are toxic when ingested over long periods of time. In a chronic feeding study, researchers feed animals a diet that contains ten to one thousand times as much of the additive as a human diet would contain. Moderately rigorous studies employ twenty-five to fifty animals of each sex of two species of rodents (usually mice and rats), as well as a smaller number of animals of another species (usually dogs). As a control, an equal or greater number of animals is fed identical food, except that the additive is left out. The animals are maintained on their diets for their entire lives: eighteen months to two years for rodents, seven years for dogs. Technicians measure the animals' weight, consumption of food, and blood composition every one or two weeks. In the more thorough studies, special tests are performed to measure the health of the liver and kidney and to detect special effects caused by a particular additive. After the animals die or are sacrificed, pathologists examine the major organs of as many animals as possible, first by eye and then microscopically, for evidence of tumors and other damage. A chemical is unacceptable as a food additive if low concentrations are harmful or if any amount causes cancer.

When we interpret the results of animal studies it is crucial that we keep one eye fixed on the effects detected and the other on the amounts of chemical that were used. Disregarding the quantity of chemical that causes a toxic effect all too commonly leads to groundless scare stories that wrongly indict innocent substances. Large enough doses of *any* chemical, including salt, sugar, and water, are injurious to animals and humans. Some substances upset the water balance, causing chronic diarrhea or increased production of urine; others taste so bad that animals would rather starve to death than eat their food; still other substances may overwork and damage the kidneys, liver, or other organs. In many cases these effects do not arise when animals consume small or moderate amounts of the chemical. Thus, just because extremely high dosages of a chemical are harmful does not mean that high levels or normal dietary levels will be harmful and that the chemical should be prohibited as a food additive.

Cancer, however, does *not* appear to be one of the effects that occurs at high dosages, but not at low dosages. If tumors are caused by feeding an animal large amounts of a chemical, we must assume that small amounts will also cause tumors, although less frequently. The 1958 food additive law recognizes the importance of eliminating carcinogens from our food supply by banning any additive that "is found to induce cancer when ingested by man or animal." This section of the Food, Drug and Cosmetic Act is known as the "Delaney Clause"; it is named after Representative James J. Delaney, Democrat of New York, whose persistence and zeal resulted in this and other consumer protection laws in the 1950s.

Before they conduct chronic feeding studies, scientists conduct shorter studies to screen out relatively toxic substances. Short-term toxicity studies usually last from one to six months. Scientists examine animals' growth, blood, urine, and how their livers and kidneys are functioning.

Once they know an animal's approximate tolerance to a chemical on the basis of short-term experiments, scientists can select the most informative dose levels for the crucial lifetime feeding studies.

Feeding studies demonstrate the effects of a chemical on the animal; biochemical experiments reveal the effects of the animal on the chemical. By learning how an animal's body handles a chemical, one can frequently estimate a substance's toxicity. For instance, some chemicals, such as carboxymethylcellulose and karaya, are not absorbed by the body and are likely to be harmless when present in moderate amounts. Additives that are normal constituents of the diet or identical to chemicals occurring in the body—or that are converted to such chemicals in the digestive tract—are generally also safe at moderate levels. Some chemicals, such as certain coal tar dyes and sodium nitrate, may themselves be harmless, but biochemical experiments can reveal that they are converted by bacteria living in the intestine or by liver enzymes into highly toxic agents. In other cases, experiments may show that potentially poisonous substances are rapidly detoxified in the liver and then harmlessly excreted. Good biochemical studies are marked by (a) the use of radioactive additive; (b) the use of animals that metabolize the chemicals as humans would; (c) follow-up studies in man; and (d) bookkeeping that accounts for all the material that was administered to the animals.

The way in which a chemical is administered to animals has great bearing on the results and interpretation of all experiments. Ordinarily, scientists add chemicals to the food or water of the experimental animals or feed the animals by means of a stomach tube. The oral route is the most meaningful way of testing prospective food additives, because additives enter the human body through the mouth. Sometimes, though, important information can be obtained only by applying the additive to the skin or injecting it under the skin (subcutaneously), into a blood vessel (intravenously), or into the body cavity (intra-peritoneally). Injecting a chemical into an animal with a syringe bypasses the intestinal bacteria, stomach acid, and digestive enzymes, any of which may alter the substance under study, and also exposes the animal to sudden, ex-ceedingly high levels of the agent. The results of this kind of experiment must be interpreted in light of metabolic studies, in case, when ingested, a substance is not ab-sorbed into the bloodstream or is converted into a second substance by the digestive system. Otherwise, the con-clusions drawn from such an experiment are likely to be erroneous.

When a chemical is injected subcutaneously, local tu-mors (fibrosarcomas) often develop at the site of in-jection. Scientists argued for years over whether these tumors are significant or are simply due to nonspecific physical irritation caused by repeated injections. Re-searchers have found that local tumors can be induced by implanting bits of inert plastic or by repeatedly in-jecting harmless nutrients (sugar, sorbic acid) under the skin. The general consensus of cancer specialists is that when tumors arise only at the site of injection the chem-ical should not be considered a carcinogen, but that when tumors develop far from the point of injection—in the liver, lung, bladder, etc.—the chemical should be viewed with suspicion and tested further in careful feeding studies. Tannin and the artificial coloring on some batches of Florida oranges are two food additives that, when in-jected, cause suspicious kinds of tumors.

In 1962 the world was shocked by an epidemic of Eu-ropean babies who were born with heartbreakingly de-formed arms and legs. Some good detective work traced the cause of deformities in four thousand babies to thalidomide, a sedative that the mothers had taken during pregnancy. The thalidomide disaster dramatized the possibility that supposedly safe chemicals in our increasingly synthetic environment may endanger the development of an unborn baby. While such tragedies are rare, birth defects are common. According to the National Foundation–March of Dimes, every year 250,000 American babies are born with such defects as a cleft palate, club foot, or an open spine. Scientific experts agree that environmental factors, such as food additives, drugs, smoking, and air and water pollution, cause a sizable percentage of the birth defects.

A person who is marked by a congenital defect must shoulder an unenviable burden through his entire life. And added to the incalculable human grief is the enormous monetary cost to the victim's family and to society in the form of doctor bills, hospital rooms, deprivation of earnings, and lack of contribution to the economy. This dollar cost runs to hundreds of thousands of dollars per person. To minimize these costs it is essential that manufacturers be required to test drugs, pesticides, food additives, and other chemicals to which masses of persons are exposed for their capacity to cause birth defects (teratogenicity).

In the 1960s the FDA began asking manufacturers of selected proposed additives to test their effects on animal reproduction. Company scientists have usually complied by integrating reproduction studies into lifetime (chronic) feeding experiments. They do this by mating pairs of rodents whose diet contains the additive; the progeny, in turn, are fed the additive and are themselves mated. A good study of this sort spans four generations and reveals a chemical's effect on fertility, lactation, and the development of the embryo.

While the multi-generation technique just described is

convenient and inexpensive, it has important shortcomings. In the first place, rarely are progeny from more than a dozen parent animals examined. This means that only the most potent teratogens could be detected. The second limitation of multigeneration studies follows from the way animals dispose of foreign chemicals. Some toxic chemicals stimulate an animal's liver to produce large amounts of detoxification enzymes, which are ordinarily present at low levels. These enzymes modify foreign substances in such ways that the substances are usually rendered less toxic and excreted more rapidly. Thus, an animal exposed for a lengthy period to a potential teratogen may possess high levels of detoxification enzymes and dispose of it much more rapidly than would an animal encountering the chemical for the first time. For this reason, in the best teratology studies animals are fed a chemical for only brief periods, usually just one to three days, during pregnancy. This method offers the greatest possibility of detecting a teratogen. As yet, few food additives have been so tested, but this situation will change in coming years as the FDA implements more stringent testing requirements. In addition, the FDA has hired private laboratories to study fifty or so widely used additives to determine if these substances cause either birth defects or mutations.

It took the thalidomide episode to drive home the importance of examining the effects of food additives on reproduction. A few farsighted scientists are now urging that we not wait for an analogous, and less easily detected, catastrophe to induce us to screen out food additives that affect our genes.

Biologists have known for several decades that radiation or chemicals can cause changes—or mutations—in the genes of plants and animals. Genes in a woman's egg cells and in a man's sperm carry, in chemically coded form, the instructions for the development of the next generation of humans. Mutations in these genes may garble the precisely inscribed instructions and cause such "mistakes" as hemophilia, extra fingers on the hand, heightened need for certain nutrients, large or small changes in intelligence, mongolism, and decreased resistance to disease. The mutations—and their effects—may persist from one generation to the next.

The importance of identifying mutagenic chemicals was cogently described by Dr. James F. Crow, chairman of the Genetics Department of the University of Wisconsin, in a National Institutes of Health Genetic Study Section report entitled "Chemical Risk to Future Generations." Crow wrote:

> Special attention should be given to the danger of very low concentrations of highly mutagenic compounds that might be introduced into foodstuffs . . . Even though the compounds may not be demonstrably mutagenic to man at the concentrations used, the total number of deleterious mutations induced in the whole population over a prolonged period of time could nevertheless be substantial. Such increase in mutation rate probably could not be detected in a short period of time by any direct observations on human beings. Protection from such effects must depend on prior identification of mutagenicity.[8]

Geneticists developed practical methods of detecting chemical mutagens in the 1960s and have tested only a few food additives. One new method, the "dominant lethal" method, is quite simple. Male mice are fed or injected with large amounts of the suspected mutagen and then mated to 10 to 50 untreated female mice. Two weeks after the mating, the researcher sacrifices the pregnant mice and counts the fetuses. If the chemical caused

[8] *Scientist and Citizen 10* 113 (June/July 1968).

a high rate of mutations in the sperm cells, the average number of live fetuses per mother would be significantly reduced. The major limitation of this technique is that only mutations that kill the developing animals can be detected.

A second way that scientists can identify mutagenic chemicals is to treat animals with large doses of chemicals and then look under the microscope for damaged chromosomes. Bone marrow, spleen, and testes contain large numbers of dividing cells and are the tissues most likely to reflect genetic damage.

Yet another way geneticists can identify mutagens is to treat cultured animal cells with a chemical and then examine the cells' chromosomes under the microscope. Broken and otherwise altered chromosomes are indicative of a mutagen. Cultured cells, however, are vastly different from animals, and just because a chemical breaks chromosomes in cultured cells indicates that it might, but does not prove that it would, cause mutations in an animal. In a live animal some chemicals would not be absorbed into the blood, others would be metabolized by intestinal bacteria or the liver, and others would not reach the germ cells.

Before they developed the techniques involving animals or animal cells, geneticists studied the effects of many chemicals, including some food additives, on pure DNA (the chemical of which genes are made), viruses, bacteria, and fungi. While these studies gave great insight into the reproduction of microbes and the biochemical basis of mutations, their relevance for man is limited, because we cannot reliably extrapolate the effect of a chemical on a microbe's genes to its effect on human genes. Some chemicals do not affect microbes but are mutagenic in humans, others cause mutations in both, and a vast number of others cause mutations in microbes but not in man. Hopefully, in the near future, the FDA will require manufacturers to test new food additives for mutagenic effects.

Testing a proposed food additive for carcinogenic, teratogenic, mutagenic, and other harmful effects, and conducting metabolic studies are major projects taking several years and costing $100,000 to $300,000. But even after this great investment of time and money in what would appear to be a thorough investigation, there remains a disconcerting variety of ways in which an apparently safe additive could cause trouble.

Weakly carcinogenic food additives[9] will probably not be detected by current testing methods. In fact, the reason we are justified in being alarmed when a scientist discovers that an additive causes cancer is because our methods are so coarse and inefficient. The primary weakness of feeding tests is that only a small number of animals is employed. In most feeding studies only twenty-five to one hundred animals are tested at each dose of chemical. (Twenty years ago scientists used only six to twenty-four animals per dose.) The results of biological experiments involving fifty or one hundred animals are inevitably blurred by spontaneously occurring tumors, statistical variations, and the premature death of some of the animals. In a typical study involving one hundred animals, at least four or five (5 percent) would have to develop cancer before the chemical being tested could, with any confidence, be called carcinogenic. Using enormous doses 100 to 1,000 times greater than dosages to which humans would be exposed partly compensates for the small number of animals, but the studies are still insufficiently sensitive. Even including the assumption that animals and humans are equally sensitive to a chemical, a

[9] A similar case could be made for weak teratogens and mutagens.

"rigorous" experiment could detect only those carcinogens that cause cancer at a rate greater than one out of 2,000 to 20,000 persons. This means that as many as 10,000 to 100,000 Americans could be stung by a "safe" food additive.

It should be emphasized that feeding studies, which because of costs inevitably involve a limited number of animals, are as sensitive as they are only because massive amounts of chemical are fed to the animals. Even strong carcinogens might escape detection if a couple of hundred rats were fed a chemical at the level at which it occurs in a human diet. Food industry spokesmen frequently argue that experiments in which only huge doses of chemical caused cancer are invalid because enormous amounts of *any* chemical will cause cancer. In fact, it is the food industry's argument that is invalid. Scientists have shown that extremely high dosages of many chemicals do not cause cancer.

In virtually all toxicologic studies only two or three species of animals are treated with the chemical. Yet we know that sensitivities to chemicals that cause birth defects and cancer vary greatly from one species to another, and it is impossible to know which species' reaction would be most similar to man's. The classical illustration of this problem is thalidomide, to which women are at least thirty to two hundred times as sensitive as rabbits, mice, rats, hamsters, and dogs.

In addition to the markedly different sensitivities of different species to a chemical, there are smaller but significant variations between strains of animals within a species. Animals used in experiments are not field mice or sewer rats recently captured in the wild, but rodents that have been highly selected and inbred over many generations in the laboratory. Laboratory animals are really subspecies and may be extremely sensitive to or uniquely resistant to a specific carcinogen or teratogen. Usually only one and rarely two strains of a species are used in a feeding study. In contrast, the entire extraordinarily diverse American population ingests food additives. Americans of African or European ancestry may react very differently to an additive than a person of Asian descent. Moreover, each subculture within the United States—blacks, adolescents, vegetarians, suburbanites, etc.—has distinctive eating habits that may increase or decrease its sensitivity or exposure to a certain chemical.

Laboratory experiments are always conducted with well-fed, pampered animals, the kind that might be most resistant to the effects of marginally toxic chemicals. Humans, on the other hand, suffer infections and diseases, may be underfed, and are frequently malnourished. These factors increase the likelihood that a foreign chemical will be harmful. In one of the rare experiments involving intentionally malnourished animals, Dr. Leo Friedman, an M.I.T. biologist who is now at the FDA, discovered that an emulsifier (polyoxyethylene-(40)-stearate) caused an abnormal increase in the number of stomach cells, which is often indicative of cancer, when rats ate food deficient in vitamin A but not when they ate a normal diet.[10] The results of this experiment should be pondered in conjunction with the results of the U. S. Department of Agriculture's 1965 dietary survey, which revealed that 36 percent of families whose annual incomes were below $3,000 and 9 percent of families whose incomes were above $10,000 had diets that contained less than two thirds of the recommended daily allowance for one or more of seven nutrients (protein, calcium, iron, vitamins A, B-1, B-2, C).

Another difference between laboratory conditions and

[10] *Fed. Proc.* 25 137 (1966).

human experience is that typical toxicity tests measure the effects on animals of one food additive in an otherwise "pure" diet. Most Americans consume a rich assortment of synthetic antioxidants, thickening agents, emulsifiers, coal tar dyes, and preservatives, as well as drugs, air pollutants, water pollutants, and pesticides. The human diet obviously abounds in opportunities for two otherwise harmless chemicals to interact and cause injury or for an organ, such as the liver or kidney, to be temporarily overworked. For instance, BHA and BHT stimulate the liver to produce detoxification enzymes which may hasten the destruction and reduce the effectiveness of certain medicines.

Still another of the inadequacies of current testing procedures is that only a limited range of effects can be detected. Birth defects, for instance, are usually looked for only in fetuses or newly born animals, although some congenital defects may become apparent only after puberty or menopause. Only a few specific kinds of mutations can be detected with contemporary techniques. Harmful effects on the senses, intelligence, or behavior may be impossible to detect in animal experiments.

The FDA reduces the hazards of food additives by including a large, arbitrary safety factor when it sets the levels allowed in food. For example, if the maximum harmless dosage of a chemical is 1 percent of an animal's diet, then the maximum amount allowed in human food would generally be set at 0.01 percent. This hundredfold margin of safety constitutes a reasonably good safeguard for persons who are especially sensitive to a chemical or whose diets contain unusually large amounts of a chemical.

At present, most testing of food additives is done by the food industry itself. The company that wishes to market a new additive and to profit from its use is in charge of the testing program to evaluate its safety. Companies either test chemicals in their own facilities or contract the work out to private testing laboratories whose financial solvency may depend upon an obsequious attitude toward food and chemical companies. One would not be surprised if greedy companies used lax experimental protocols or overlooked an occasional tumor. While the fear of lawsuits prevents most companies from engaging in malicious deceptions, industry scientists might unthinkingly interpret ambiguous results in a way favorable to their company. Concerned scientists and lawyers have suggested that this built-in conflict of interest be eliminated by inserting a government or other disinterested middleman between the company that proposes a food additive and the laboratory that tests it. This system would also enable testing laboratories to be truly independent.

As scientific advances are made and new techniques developed, animal experiments will be increasingly predictive of the effects of food additives on human beings. Meanwhile, we should recognize the limitations of "thorough" studies and should evaluate with great conservatism and some skepticism the experiments upon which is based the safety of chemicals to which two hundred million persons may be exposed.

Chapter 2
A Close-up Look at Food Additives

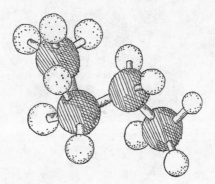

INTRODUCTION

> "Out of this world! Amazing! Stupendous!
> Unbelievable . . . Virtually no limit to the
> natural foods that now can be duplicated with
> special characteristics that make them better
> suited to today's modern needs. And CPC
> International, with its related affiliates, has
> virtually everything you need to enter this
> growing market . . . the carbohydrate, the fats,
> the oils, the flavors—and the expertise. All you
> need is the protein—and the will."
>
> Advertisement, CPC International
> From *Food Product Development*, October 1971

The uses, abuses, and safety of the most widely used food
additives are described in the following pages. Substances
are listed in alphabetical order under their usual name.
Some compounds, such as iodine and nitric acid, are
most frequently used as sodium, potassium, or calcium
salts and are therefore listed as potassium iodide, sodium
nitrate, etc. Consult the index when you have difficulty
finding a certain additive.

References to a few of the more informative, recent, or
interesting publications are given for most additives. Abbreviations used in the references (except for scientific
journals) are deciphered in the bibliography. Additional
articles may be located by referring to papers cited in
these publications. The most recent scientific and medical
papers are best found by using *Chemical Abstracts, Index
Medicus,* and *Science Citation Index.* Actions taken by
the Food and Drug Administration, as well as its trials and
tribulations, are reported in the pages of the *Washington
Post* and in *Food Chemical News,* a trade publication.

The cryptic numbers and letters at the end of most lists of references are a guide to the regulatory status of the additive. The combination beginning with 21 CFR, such as 21 CFR 121.1180, refers to the official government listing of food additives in Title 21 of the Code of Federal Regulations. Individual regulations list the foods in which the additive may be used and at what levels. The safety of most of these chemicals is supported by varying amounts of scientific testing. "GRAS" indicates that the FDA has deemed the additive to be "Generally Recognized As Safe." That a chemical is generally recognized as safe does not mean that it *is* safe, as discussed in the Glossary (see "GRAS," Appendix 4). (Cyclamate used to be GRAS.) Many GRAS chemicals are listed in 21 CFR 121.101(d-g), 21 CFR 121.1163-4 (flavorings), and in Appendix 2 of this book.

For explanations of scientific terms and food additive jargon, turn to the Glossary (Appendix 4). The testing procedures referred to in the discussions of most additives are fully discussed in Chapter 1 and summarized under "toxicity tests" in the Glossary.

Drawings of the chemical structures of many of the additives are presented in Appendix 3.

Acetic Acid   GRAS

Acetic acid is the substance that gives vinegar its sharp taste and odor. Manufacturers use it to preserve, flavor, or acidify foods.

Almost all of the billion pounds-plus of acetic acid produced each year in the United States is produced chemically from alcohol and acetaldehyde. Much of this is used in manufacturing plastic; the food industry uses only a trivial amount. Vinegar is produced by the bacterial fermentation of low-grade fruit and fruit by-products (apple cores, peels, etc.). Bacteria convert carbohydrate in the fruit first to alcohol and then to acetic acid.

Alcohol can also be converted to acetic acid by oxygen in air; acid produced in this way accounts for sour wine. Historical records show that mankind has been afflicted with this contretemps for at least five thousand years.

Ref.: ECT 8 386; 21 254.

Acetone Peroxide

Bakers have known for centuries that you cannot make good baked goods from freshly milled flour. Dough made from such flour lacks the strength and resilience needed to trap the gas bubbles that are produced by yeast, leavening, or whipping and that create fluffy, tender bread and cake. Until the twentieth century, bakers stored flour for months before using it, to let the oxygen in the air gradually condition the flour. However, at the same time the flour was aging, vermin would defile or devour part of it and another portion would spoil. To eliminate the costly and wasteful storage period, food chemists devised chemical methods that age flour almost instantaneously. Acetone peroxide is one of the newer and less frequently used of the chemical aging agents. Like several other flour improvers, acetone peroxide bleaches the flour white in addition to aging it. This powerful oxidizing agent is a solid at room temperature and is used at levels of 5 to 120 parts per million (ppm), depending upon the flour.

Acetone peroxide has been tested on animals but not extensively. In one study, scientists fed four generations of rats bread or flour that had been treated with 450 ppm of this chemical. The treated food comprised 70 percent of the rats' diet. Pathologists did not detect tumors, birth defects, infertility, or tissue damage, but they did not examine the lungs or urinary bladder, two organs that frequently develop tumors. A one-year study on a small number of dogs did not reveal any adverse effects. A lifetime feeding test involving at least one additional species, as well as careful genetic and teratogenic studies, must be conducted before we can consider acetone peroxide adequately tested.

Ref.: FAO(40A)-117; *Bakers' Digest 36* 50 (1962); *Fd. Cos. Tox. 5* 309 (1967); 21 CFR 121.1023.

Adipic Acid GRAS

Many manufacturers use adipic acid as the acid in bottled drinks and throat lozenges. Because this additive has little tendency to pick up moisture, it is frequently used to supply tartness to powdered products, such as gelatin desserts and fruit-flavored drinks.

Adipic acid is occasionally added to edible oils to prevent them from going rancid. The acid's two negatively charged groups attract and trap positively charged metal ions, which may be present in the oil. Free metal ions promote chemical reactions that cause noxious odors and off-colors to develop in oils.

Rats, and presumably humans, metabolize adipic acid without any difficulty. One small-scale lifetime feeding study on rats indicated that moderate doses (1 percent of diet) are harmless.

Ref.: FAO(40A)-129; C&EN-117; CUFP-67.

Agar GRAS

Agar is a tasteless, odorless carbohydrate that is extracted from seaweed. Food manufacturers seldom use agar, but it is of great importance to the biomedical profession because it forms a stable gel that provides a good medium for growing cultures of microorganisms.

The natural source of agar is red algae, which grow off the coasts of Japan, Spain, and southern California, ten to forty feet below the ocean's surface. The seaweed is harvested by hand by divers and then boiled in water.

The boiling extracts the agar from the plant and when the water cools an agar gel forms. The gel is dried and the agar ground to a powder. This is essentially the same purification scheme as was used in 1658 by the Japanese innkeeper who first discovered the gel-able constituent of seaweed.

The most important use of agar is as a solid support for growing bacteria. The idea of using agar in bacteriology was conceived in the nineteenth century by a German housewife. Her husband, a physician, had success with the technique and passed the idea along to Robert Koch, who proceeded to use it in 1882 to discover the bacterium that causes tuberculosis. Since that time agar gels have been standard equipment in microbiology laboratories.

Gels made from agar are tough and brittle compared to gels made of gelatin or carrageenan and do not "melt in the mouth," so they find little use in foods. However, agar is used in food at concentrations too low to form a gel. Its major use in the United States is in baking, where it prevents icings on cakes and cupcakes from drying out. On occasion it serves as a thickening agent in ice cream, jam, and whipped cream.

Agar is nondigestible and swells greatly in water, suggesting another of its uses: as a bulk-type laxative.

Long-term animal feeding studies have not been done, but traditional use in food, short-term animal experiments, and medicinal experience in man all point to agar's lack of toxicity.

Ref.: FAO(35)-64; *ECT 17* 763.

Alginate[1]   Propylene Glycol Alginate   GRAS

Algin was first isolated from brown algae in 1881 in England. In the seaweed, algin constitutes an important part

---

[1] The ammonium, calcium, potassium, and sodium salts of algin.

of the cell wall; as a food additive it acts as a thickening and stabilizing agent. The most important commercial source of algin is giant kelp, a seaweed that is "farmed" off the southern California coast.

Algin solutions can be converted to gels by adding calcium. These chemically set gels are extremely stable, and while they do not possess that "melt in your mouth" texture, they help prevent jelly in pastries from oozing all over the oven. Industry also uses algin to help maintain the desired texture in ice cream, cheese, candy, pressure-dispensed whipped cream, yogurt, canned frosting, and other factory-made foods.

Short-term feeding studies on animals have shown that alginate is not overtly toxic. Alginate is not absorbed by the body, thus it probably does not cause cancer or birth defects, but until appropriate experiments are done, this chemical (and any impurities in it) cannot be pronounced safe. Like most other thickening agents, alginate binds a great deal of water in the gastro-intestinal tract and could inhibit the absorption of nutrients. This possibility needs to be studied.

In the early and middle 1960s Americans worried a great deal about radioactive fallout contaminating our food supply. Strontium 90 was particularly worrisome because it behaved much like calcium: it was present in milk and was deposited in bones and teeth. Scientists who were pessimistic—but perhaps realistic—enough to assume that radioactive fallout was going to be an unavoidable ingredient of modern life sought ways to minimize the hazard of fallout. Doctors working with alginate made the surprising discovery that the seaweed derivative had a much greater affinity for strontium than for other ions, including strontium's close relative calcium. When 1 percent algin was added to milk, the body's absorption of strontium was reduced by 75 percent while calcium absorption was decreased only 35 percent. Fortunately, the U.S.-Soviet ban on above-ground nuclear tests made further development of this discovery unnecessary.

Alginate cannot be used as a thickener in all foods because it precipitates out of acidic foods, such as soft drinks and salad dressing. Chemists discovered that they could overcome this limitation by reacting alginate with propylene oxide. The derivative formed, propylene glycol alginate, is stable in acid and has good thickening ability. You will find propylene glycol alginate used as a thickener or stabilizer in ice cream and other frozen desserts (up to 0.5 percent), cheese spreads (up to 0.8 percent), salad dressings (up to 0.75 percent), and beverages (up to 0.02 percent).

Propylene glycol alginate is exceptionally proficient at stabilizing the foam in beer. The reason why such an additive is needed in many American beers was described in *American Brewer,* the beer industry's magazine:

> When beer was made with more malts and hops than is the practice today, stability of foam was rarely a problem. Today's lighter-type lagers and ales, however, made with higher percentages of adjuncts and only mildly hopped, require assistance in maintaining an expected head of foam.

The next time you see a beer's foamy head touted in a television commercial, remember the chemical additive that deserves the credit.

Lifetime feeding studies involving the modified alginate have not been conducted, but a biochemical study is in progress. Until the results of these tests are in and reproduction studies are conducted, propylene glycol alginate cannot be declared free from hazard.

Ref.: FAO(35)-68; FAO(46A)-135; *Can. Med. Asso. J. 99* 986 (1968); *Am. Brewer* p. 22 (December 1963); 21 CFR 121.1015 (propylene glycol alginate).

Alpha Tocopherol   (Vitamin E)   GRAS

Tocopherol is a vitamin that is abundant in wheat and rice germ and vegetable oils. Most Americans' diets are severely deficient in vitamin E because little or none is present in white flour, enriched white bread, and enriched white rice. According to Dr. A. L. Tappel, a biochemist at the University of California at Davis, the average American gets less than one fourth of the daily recommended dose. Vegetable oil remains a good source of this vitamin, although approximately 25 percent is lost during typical commercial processing. Both in nature and as a food additive, tocopherol acts as an antioxidant, preventing fats and oils from going rancid. See also "butylated hydroxyanisole" and "vegetable oil."

Ref.: Am. J. Clin. Nutr. 17 1 (1965); *Vitamin E Content of Foods and Feeds for Human and Animal Consumption,* Dicks, M. W., U. Wyoming, Laramie, Bulletin 435 (December 1965); *Let's Eat Right to Keep Fit,* Davis, A., New American Library, N.Y. (1970).

Ammoniated Glycyrrhizin   GRAS

Licorice can be hazardous to your health. The medical literature describes twenty or thirty persons whose hearts began to fail when they ate several ounces of licorice candy a day for extended periods of time. One healthy fifty-three-year-old man developed congestive heart failure, hypertension, fatigue, edema, and headaches when he consumed twenty-four ounces of licorice in nine days. Symptoms disappear a week or two after the patients eliminate licorice from their diet. The toxic agent in the candy is monoammonium glycyrrhizin, one of licorice's principal flavor components (0.5 percent).

Ammoniated glycyrrhizin has a variety of physiological effects: it raises the blood pressure, alleviates stomach ulcers, and reduces the toxicity of strychnine, carbolic acid, and diphtheria toxin. Food companies use it in licorice flavoring and as a component of root beer and wintergreen flavorings in beverages (50 ppm), candy (5 to 60 ppm), and baked goods (5 ppm). Glycyrrhizin is one of the sweetest natural substances known—one hundred times as sweet as sugar—so it is particularly useful when sweetness and licorice taste are both required.

Ammoniated glycyrrhizin is a potent drug and should be tested for its ability to cause cancer, birth defects, and infertility. Its effect on children, who may consume enormous amounts of licorice, must be thoroughly investigated. As a minimal safety measure, packages of licorice candy should carry warnings against eating excessive amounts.

Ref.: J. Am. Med. Asso. 205 492 (1968), 213 1343 (1970); J. Pharm. Pharmacol. 15 500 (1963); Chem. Abst. 63 1017c, 11933g.

Amylases   GRAS

Plant seeds contain rich reserves of nutrients that supply growing power until the plants can produce their own nutrients by photosynthesis. In cereal grains, energy is stored in the form of starch, which is broken down to sugar as the seeds germinate. The enzymes that convert starch to sugar are called amylases.

Plants contain two different kinds of amylases. Alphaamylase attacks internal parts of starch molecules, creating new tips, while beta-amylase nibbles away at the tips. The large fragments of starch created by alpha-amylase are called dextrin. The small pieces generated by betaamylase are called maltose, a close relative of sugar (sucrose). In addition to occurring in plant seeds, alphaamylase is present in saliva, pancreatic juice, and microorganisms. Amylases, like all enzymes, are proteins and are nutritious as well as safe.

Bakers add alpha-amylase to bread dough to supplement the small amount that the wheat grain naturally contains. The amylases convert a small fraction of the starch
to sugar and dextrin as the oven's heat disrupts starch
granules. The enzymatic action stops as the rising tem-

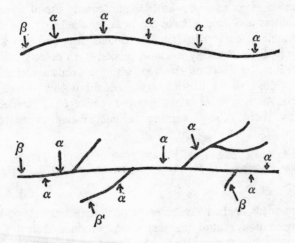

Amylase digestion of linear and branched
starch molecules.

perature in the oven destroys the enzymes. The sugar
that the amylases produce provides energy for the fer-
menting yeast and also makes for better-tasting, better-
toasting bread. The conversion of a portion of the starch
to dextrin improves the dough's consistency and the
bread's keeping quality.

Brewers use malt, which is simply germinated grain, to
convert starch to sugars. Yeast, then, converts the sugar
to alcohol. Malt is used instead of ungerminated grain

because as a grain of wheat germinates, it develops huge quantities of alpha-amylase.

In another application amylases obtained from microorganisms are used to convert cornstarch to corn syrup or to dextrose.

Ref.: CRC-69-76.

Arabinogalactan (Larch gum)

Most natural compounds that manufacturers use to artificially thicken food are obtained from seaweed or from plants that grow in hot, arid climates. Larch gum is an exception. Producers obtain it from the Western Larch tree which is abundant in the northwestern United States. The gum can be collected by tapping trees, but more commonly it is extracted from wood chips because it is very soluble in water.

Like gum arabic, another very soluble gum, arabinogalactan forms thick solutions only at concentrations above 10 to 20 percent. Most other gums form thick solutions at concentrations of 1 percent or less.

Pure arabinogalactan is a carbohydrate consisting of galactose and arabinose in a ratio of six parts to one. Commercial preparations, however, contain significant amounts of tannic acid and other impurities.

As is true for most thickening agents obtained from natural sources, larch gum has been poorly tested. Small-scale, six-month studies on rats and dogs failed to reveal any toxic effects, but lifetime feeding, reproduction, and other tests need to be done before arabinogalactan can be deemed safe. The gum itself is probably harmless, but the tannic acid impurities raise questions (see "tannin").

Ref.: TAPPI 46 544 (1963); *Chem. Abst. 62* 13759a, *66* 77206, *70* 106797; personal communication from FDA (summary of testing data); 21 CFR 121.1174.

ARTIFICIAL COLORING[2]

I. Coal Tar Dyes

Natural foods contain pigments that do wonders for the attractiveness of a meal. Vegetables, fruits, meat, grains, and beverages sparkle with almost every color in the rainbow. However, with the advent of factory-made foods, especially since World War II, many of our meals might be drab affairs were it not for the generous use of food coloring. Imagine a meal with colorless cherry soda, fat-gray frankfurters, colorless frozen lemon cream pies, pale raspberry sherbet, and off-white imitation pudding.

For thousands of years cooks livened up their dishes with colorings derived from seeds, fruits, and herbs. While these natural products are still used as coloring, they have drawbacks such as fluctuations of supply and price, low coloring power, fading of color, and high cost. All of these difficulties were overcome—but not without the introduction of safety problems—when scientists discovered how to synthesize colored compounds from, of all things, coal.

Coal, when heated in the absence of air, is converted to coke (impure carbon), coal gas, and coal tar. The coal tar, a viscous black liquid, is a mixture of many organic compounds. In the 1800s, English and German chemists learned that they could produce intensely colored substances when they purified some of these compounds and reacted them with other chemicals. These synthetic sub-

---

[2] General references on artificial colorings: FAO(38B); FAO(46A); 21 CFR sections 8 and 9; CRC-Ch. 1; *Fd. Cos. Tox. 4* 186 (1966); *Food Tech. 22* 946 (1968) has tables that list the quantities of dyes used by various sectors of the food industry.

stances are known as coal tar dyes and are used in great quantity by the food, fabric, and cosmetic industries.

Americans are consuming coal tar dyes at an increasingly rapid rate. According to the FDA, in 1940 251,000 pounds were certified[3] for use in foods (or drugs or cosmetics). In 1950 and 1960 the corresponding figures were 1,474,000 and 2,159,000 pounds. In 1970, slightly over 3,735,000 pounds were certified by government inspectors. Thus, usage has increased fifteenfold in thirty years. Over 95 percent of these dyes are used in foods, particularly beverages, candy, ice cream, dessert powder, baked goods, and sausages. Levels used range from approximately ten parts to five hundred parts per million, the lower levels being used in liquid foods (beverages, gelatin desserts) and the higher levels in solid foods (pet food, breakfast cereals).

Coal tar dyes frequently provide the only color in factory-made foods. In no case does artificial coloring add to the nutritive value of the food which it colors. Rather, the presence of coloring in a food generally signals a deficiency or absence of natural and often nutritious ingredients whose colors the synthetic dyes seek to imitate.

A small, but good, example of how artificial colorings can be used to deceive consumers was recently furnished by the L. A. King Food Products Company, Denver, Colorado. *FDA Papers* (December 1970 and April 1971) reported that government inspectors caught that company adding two yellow food colorings (Yellow 5 and Yellow 6) to its "Grandma's Fresh Frozen Egg Noodles" to make the noodles look like they contained more egg than they did. Poor Grandma.

---

[3] "Certified" means that a batch of coloring has been inspected and contains less than a certain level of impurities; certification does *not* mean that adequate toxicology studies have been conducted.

Manufacturers often use artificial coloring in candy, breakfast cereal, and pet foods to create variety or eye appeal rather than to hide the absence of an ingredient or mimic a natural food. These rather frivolous uses of chemicals whose safety is not of the highest repute are ill-considered and should be discontinued.

Rarely will a major ice cream, pet food, or other manufacturer stop using artificial coloring in a product that all other manufacturers color. A lone conscientious company in a crowded marketplace would risk being placed at a serious competitive disadvantage. Either trade associations or the FDA will have to make industry-wide rulings. Persistent citizen pressure on the FDA and elected officials might be effective in this area. Meanwhile, concerned consumers can avoid most artificial coloring by reading labels and eating good, fresh food.

At present nine dyes may be used in any food (Blues 1 and 2, Green 3, Reds 2, 3, and 40, Violet 1, and Yellows 5 and 6) and three are limited to one particular product each. Red 4, formerly one of the most important dyes, was restricted in 1965 to maraschino cherries when it was discovered that high dosages harmed the adrenal glands of dogs. Orange B has been used since 1966 to color frankfurters, and Citrus Red 2 is used only to color the skin of oranges. Table I shows the extent to which the dyes were used in various categories of food in the first nine months of 1967.

People have been concerned about the possible toxicity of coal tar dyes ever since they were introduced in the second half of the nineteenth century. These complex synthetic compounds were initially suspect because they are found nowhere in nature, and it was not clear how humans would react to them. In many cases, intuition has been backed up by scientific experiments, which showed that some dyes are potent carcinogens. Repeatedly, col-

Table I Sales of artificial colors*

Category	FD&C Blue No.1	FD&C Blue No.2	FD&C Green No.3	Orange B	FD&C Red No.2**	FD&C Red No.3	FD&C Red No.4	FD&C Violet No.1	FD&C Yellow No.5	FD&C Yellow No.6	Total
Candy, confections	6,632	2,499	124	0	67,637	11,665	0	1,459	59,903	52,770	202,689
Beverages	15,800	2,375	301	0	282,695	1,056	0	985	78,933	181,292	563,437
Dessert powders	3,270	1,659	14	0	62,363	8,616	0	0	59,961	51,622	187,505
Cereals	843	99	0	0	15,558	1,421	0	0	52,496	35,464	105,881
Maraschino cherries	597	0	98	0	8,104	3,469	11,308	0	5,644	4,830	34,050
Pet food	1,473	6,764	0	0	67,058	1,023	0	1,278	101,743	23,226	202,565
Bakery goods	3,680	673	7	0	43,522	9,560	0	369	77,885	42,203	177,899
Ice cream, sherbet, dairy products	2,599	179	7	0	29,697	621	0	45	35,048	23,868	92,064
Sausage	647	0	0	16,890	36,084	4,970	0	0	6,502	99,605	164,698
Snack foods	305	0	0	0	3,623	766	0	2	18,456	11,409	34,561
Meat inks	11	0	0	0	12	10	0	2,223	15	0	2,271
Miscellaneous	5,345	1,990	1,298	0	46,219	18,200	398	1,134	44,841	29,134	148,559
Subtotal (food use)	41,202	16,238	1,849	16,890	662,572	61,377	11,706	7,495	541,427	555,423	1,916,179
Pharmaceutical	3,250	593	220	0	21,179	12,168	1,186	347	17,275	15,938	72,156
Cosmetics	397	30	27	0	3,417	903	630	96	3,125	2,148	10,773
Total	44,849	16,861	2,096	16,890	687,168	74,448	13,522	7,938	561,827	573,509	1,999,108

Ref.: Food Colors, Committee on Food Protection, National Research Council, Washington, D.C., 1971.
* Figures represent sales in pounds for the first nine months of 1967 and do not include exports or sales to jobbers and other manufacturers.
** Amaranth.

ors approved for use in food have been shown to be toxic or carcinogenic and have been banned. The history of approved dyes reads like the guest register in a hotel for transients (see Appendix 1).

In view of the repeated hazards associated with coal tar dyes, one might assume that the dyes now in use have been thoroughly tested. That is not the case. Adequate lifetime feeding studies (two rodent species, forty or more animals per dosage level), which must be done to reveal carcinogenic and cumulative effects, have been conducted for only five dyes (Yellow 6, Red 4, Green 3, Orange B, and Citrus Red 2). Lifetime studies on dogs have been performed only with Orange B, Red 2, Red 4, and Yellow 6. Not one of the coal tar dyes used in food has been adequately studied for causation of mutations. Only one dye, Red 40, has been tested for the ability to cause birth defects.

In September 1971 the FDA gave manufacturers one year to evaluate the possibility that color additives cause birth defects and two years to evaluate their effect on other aspects of animal reproduction. If tests are not submitted, or if tests indicate that a chemical is unsafe, the FDA announced that it may ban the colorings. It will be interesting to see if the FDA sticks to its guns or grants perpetual extensions of the deadlines.

The notion of verifying the safety of chemicals used in food is hardly a radical idea and is incorporated into this country's laws. The 1960 food coloring law states that new coloring agents must be proven safe before they may be used in food. Prior to 1960 the government had to prove that a coloring was actually dangerous before it could ban it from our food. For color additives already in use at the time the 1960 law was passed, provisional approval was granted for a period of two and a half years (until January 1963), at the end of which time safety experiments were to be completed or use of the additive discontinued. Unfortunately the law also included a clause which permitted the FDA to extend the two-and-a-half-year deadline for an indefinite period. And this is exactly what has happened for many of the dyes in use in 1960. Hopefully, now that Commissioner Charles Edwards of the FDA has asked manufacturers to test the effect of coloring on animal reproduction, the procrastination will soon end.

The safety of Violet 1 has already fallen under question. The dye is used to stamp the Department of Agriculture's inspection symbol on meat and also to color candy, beverages, and pet food. The FDA certified almost 67,000 pounds of this dye in 1972.[4] Consumers may ingest a small amount of the dye when they eat a steak or roast. Workers in packinghouses may be exposed to much greater amounts. The men who imprint USDA's mark on carcasses frequently get the dye on their hands; occasionally large amounts drip down their hands and arms. One Department of Agriculture meat inspector in North Carolina observed that "the dye is pretty hard to get off. Once you get it on your skin, you almost have to wear it off."

FDA expedited approval of Violet 1 as a food coloring at the behest of Malcolm Carroll, the former manager of Allied Chemical's certified color division. Carroll pleaded at a hearing that his company needed quick approval so they could gear up for the 1951 Easter egg coloring season and for fall 1951 candy orders.

Scientists have conducted three chronic feeding studies on rats. The earliest two studies, using male rats only, were done by Dr. Lloyd Hazleton, then at George Wash-

[4] *Ref.:* FAO(38B)-107; Fed. Sec. Agency (FDA) Docket No. FDA-58 (February 1950); *J. Pharm. Pharmacol. 14* 378 (1962); *Fd. Cos. Tox. 2* 345 (1964).

ington University, and by Dr. O. Garth Fitzhugh at the
FDA. Hazleton's study involved twenty rats per dosage
level, while Fitzhugh's involved only fifteen. Neither study
indicated that Violet 1 was hazardous. The third study,
published in 1962 by Dr. W. A. Mannell and his colleagues at Canada's Department of National Health and
Welfare, involved more rats, rats of both sexes, and higher
dosages than were used by Fitzhugh. This study indicated
that the dye caused cancer. Of thirty rats (fifteen of
each sex) fed 3 percent dye in their food for seventy-five
weeks, five developed malignant tumors. Three of the
tumors were skin tumors. Four of the five tumors occurred
in females. Only one out of thirty untreated rats developed
a tumor.

The FDA discounts the Canadian study because there
is no proof that the Canadian dye fit precisely U.S. specifications for FD&C Violet 1. (The entire batch of dye has
been used up, the manufacturer has gone out of business,
and all records have been lost.) Yet the FDA has no
positive evidence that the dye did, in fact, differ significantly from U.S. specifications.

In tests conducted by FDA toxicologists on dogs, diets
containing 0.5 to 2 percent dye caused skin lesions. There
has been a difference of opinion on whether these lesions are indicative of cancer, but most pathologists are
doubtful. Cancerous or not, Dr. Kent J. Davis who was
formerly with the FDA (and now works for the U. S.
Environmental Protection Agency) has recommended that
because the dye apparently causes skin lesions it should
be banned from cosmetics and drugs that are applied to
the skin.

At its December 1964 meeting, the Expert Committee
on Food Additives, which is cosponsored by the Food
and Agriculture Organization and the World Health Organization (FAO/WHO) of the United Nations, cate-

gorized Violet 1 as a dye "for which the available data are inadequate for evaluation but indicate the possibility of harmful effects." Dr. Jack Dacre, a New Zealand biochemist whose field of interest is coal tar dyes, wrote that this coloring "has possible harmful effects and should not be allowed to be used in food."

In early 1971, nine years after Violet 1 was first suspected of being carcinogenic, the FDA asked the National Academy of Sciences to convene a group of nongovernment scientists to evaluate all available studies. The committee met in Washington on September 1, 1971 (the author's request to sit in on the meeting was rejected by the chairman), and filed its final report in November. The committee's conclusions and recommendations were finally made public in March 1972, more than a year after the committee was first announced.

The NAS committee dismissed out of hand the Canadian study that indicated that Violet 1 caused cancer. The committee declared the dye safe, but did recommend that a lifetime feeding study be conducted on dogs. The net effect of their report will be to postpone the permanent acceptance or banning of this coloring for at least eight years.

The chronic toxicity tests done on Violet 1 do not meet the moderate standards suggested by the Mrak Commission (Report of the Secretary's Commission on Pesticides and Their Relationship to Environmental Health, 1969), the Joint Expert Committee on Food Additives of the FAO/WHO (fifth report, 1961), or the National Academy of Sciences–National Research Council (Food Protection Committee, "Evaluating the Safety of Food Chemicals," 1970). According to these committees, minimum requirements for lifetime feeding cancer studies include at least twenty to twenty-five animals of each sex of two or more species for several dosages tested. In addition, animals dying in the course of the experiments should be subjected to detailed examination. The studies involving Violet 1 are bad jokes when compared with these standards. The National Academy of Sciences committee that evaluated Violet 1 showed disdain for the public's health and ignorance of food additive safety criteria when it approved the unrestricted use of this dye.

Violet 1 is the most poorly tested food color:

—one study on rats indicated that the dye caused cancer.

—a study on dogs indicated that it might cause skin lesions.

—it has not been tested for causation of birth defects.

—if the FDA rejects the Canadian study, then it can be said that the dye has never been tested on female rodents, a *sine qua non* of any decent investigation.

It is ironic that the dye used to indicate that meat is safe may itself be hazardous. I recommend strongly that Violet 1 be banned.

Citrus Red 2 is another questionable coal tar dye.[5] Florida orange growers use the dye, mainly from October through December, to cover up the mottled green color on oranges, tangelos, and temple oranges. Without the added color many Florida growers fear they will lose sales to California oranges, which have a natural, uniformly orange color. An orange may contain dye at levels up to two parts per million. The dye does not penetrate into the pulp of the fruit.

Feeding studies on mice and rats have indicated that small amounts of the dye are harmless and do not cause cancer. However, four other lines of evidence suggest that the dye might indeed be a weak carcinogen:

[5] *Ref.:* FAO(46A)-30; *J. Pharm. Exp. Ther. 134* 100 (1961); *Proc. Univ. Otago Med. Sch. 43* 31 (1965); *Fd. Cos. Tox. 4* 455, 493 (1966); *Br. J. Cancer 22* 825 (1968); *New Zealand Med. J. 73* 74 (1971).

—when the dye was mixed with cholesterol and implanted in the urinary bladders of mice, 14.5 percent of the animals developed tumors; cholesterol without dye caused tumors in 4.5 percent of the mice.

—when the dye was injected under the skin of mice, malignant tumors developed in the lungs.[6]

—the liver converts the dye to less noxious substances, which then pass into the urine; one of the intermediates in the conversion process is 1-amino-2-naphthol; there is evidence that this chemical causes cancer.

—the walls of the urinary bladders were markedly thickened in animals whose diet contained the dye.

On the basis of this array of circumstantial evidence the FAO/WHO Expert Committee issued the following warning at its 1969 meeting:

Citrus Red 2 has been shown to have carcinogenic activity and the toxicological data available were inadequate to allow the determination of a safe limit, the Committee therefore recommends that it should not be used as a food color.

American consumers are endangered if they eat or suck the peel of treated oranges or use the peel in marmalade. Unfortunately, oranges are no longer individually stamped "color added" so it is difficult to identify treated oranges. When oranges are not stamped, supermarkets are supposed to post signs saying "artificially colored oranges." FDA administrators in Washington admit that grocers from one end of the country to the other are ignoring FDA's regulations, but plead that they have too little manpower to stop this "minor" infraction. Workers who produce Citrus Red 2 and who dye

[6] Tumors caused by injected chemicals are considered significant only when they appear at sites distant from the injection site, as was the case for Citrus Red 2.

the oranges may ingest or inhale relatively large amounts of the chemical. They would be exposed to a proportionately greater hazard than the average consumer.

Fortunately a decreasing number of shippers are dyeing their oranges. The percentage of orange shipments that were dyed decreased from 72 percent in 1947 to 36 percent in 1972. One reason for this decrease, according to Mr. H. M. Riley, a U. S. Department of Agriculture inspector stationed in Florida, is that Florida oranges compete very little with California oranges. Florida growers control the eastern market, while California growers control the western and much of the midwestern markets. Thus, consumers can rarely compare the appearances of the two kinds. Another reason for the decreased use of the dye may be that at least eleven states and Canada have outlawed the sale of artificially colored oranges.

The orange growers' diminishing use of Citrus Red 2 is a move in the right direction, and a move that the Food and Drug Administration should endorse by banishing this suspicious chemical to the scrap heap of history. The American people ingest enough carcinogens already and should not be exposed to an extra dose on their oranges.

The identification of coal tar dyes on food labels is as inadequate as the testing the chemicals have undergone. The presence of coloring in butter, cheese, and ice cream need not be specified at all. In other foods, colorings are never identified specifically as Violet 1, Green 3, etc., but only as "artificial coloring." This is obviously a great inconvenience to persons—such as sufferers of severe allergies—who must avoid only one or two certain colorings. Because of the vague labeling, a person who is allergic to one food coloring may have to avoid all foods containing any artificial coloring. Allergists (and their patients) have urged that all food additives including artificial colors be specifically identified on all food products. Dr. F. H.

Chafee, a physician at the Rhode Island Hospital in Providence, pleaded for complete labeling in a paper in *Journal of Allergy* (*40* 65 [1967]). According to Dr. Chafee:

> We have had cooperation from the medical departments of the companies concerned when we have asked for the dye in a specific product and have given the reasons for our request. Even so, it has been a time-consuming procedure. It is difficult for the average physician, and nigh impossible for the patient, to obtain this information. It would be far easier for the physician, let alone the patient sensitive to these chemicals, if the FDA were to require listing of the dyes on the package . . . Required listing of these dyes on drug and food packages might be life-saving.

One simple way of indicating which colorings are used in a particular product would be to assign each of the coal tar dyes a code letter or number. Red 3 could be R3, Green 3 could be G3, etc. The label listing for artificial coloring would read: ". . . artificial coloring (R3, G3, Y5)." The identity of the chemicals for which the letters stand would be public information.

Other Coal Tar Dyes

Blue 1 (Brilliant Blue FCF). American and Canadian biologists have conducted lifetime feeding and biochemical studies of this dye on rats. The dye was poorly absorbed in the intestine and appeared to be harmless. Studies are needed on large numbers of animals of other species. FDA inspectors certified more than 83,300 pounds of Blue 1 in fiscal year 1970.[7]

Blue 2 (Indigotine). Lifetime feeding studies have been conducted on rats, and good short-term studies have been conducted on pigs (whose physiology resembles man's) and dogs. In rats, Blue 2 is partially absorbed in the intestine and metabolized, causing no detectable harm.

[7] *Ref.:* FAO(38B)-27; FAO(46A)-24; *Tox. Appl. Pharm.* 8 29 (1966); *J. Pharm. Exp. Ther.* 114 38 (1955), *163* 222 (1968).

Dogs whose food contained 2 percent dye appeared to be more sensitive to fatal virus diseases; no other effects were noted. This dye needs further study. In fiscal year 1970 the FDA certified 39,974 pounds of Blue 2.[8]

Green 3 (Fast Green FCF). Dr. O. Garth Fitzhugh, an FDA pathologist, directed lifetime feeding studies on rats and mice; he found no evidence of carcinogenicity. Green 3 is poorly absorbed by rats and dogs, but more detailed metabolic studies are needed. In fiscal year 1970 FDA inspectors certified 5,005 pounds of Green 3.[9]

Orange B. Meat packers in certain areas of the country use Orange B to color the outer surface of frankfurters (up to 150 ppm). The advantage of Orange B over other dyes is that it remains at the surface of the sausage, so less is required. Dr. John Doull, then at the University of Chicago, conducted lifetime feeding studies on dogs, rats, and two strains of mice (one a cancer-susceptible strain). These studies were well designed but poorly executed, because the tissues from fewer than thirteen of the one hundred rodents per dosage level were examined microscopically. More thorough rodent studies are needed. Aside from liver nodules formed in dogs whose diet contained 2 percent or more dye (the meaning of the nodules is unknown), Doull and his associates did not detect any adverse effects. In fiscal year 1970 the FDA certified 34,000 pounds of Orange B.

Doull's studies are inadequate to substantiate the safety of Orange B. In addition, the chemical structure of Orange B is closely related to that of Red 2, which causes birth defects in animals (see p. 65). There is no good reason why we should be consuming seventeen tons of this dye

[8] *Ref.:* FAO(38B)-63; FAO(46A)-43; *J. Pharm. Exp. Ther.* *154* 38A (1966); *Tox. Appl. Pharm.* 8 29 (1966).
[9] *Ref.:* FAO(38B)-56; FAO(46A)-38; *J. Pharm. Exp. Ther.* *114* 38 (1955); *Fd. Cos. Tox.* 4 389 (1966).

in our frankfurters every year. Coating hot dogs with dye adds a totally unnecessary hazard to our food supply and should be prohibited throughout the country.[10]

Red 2 (Amaranth). In September 1971 the FDA made the startling announcement that its own study had verified the results of a Russian study indicating that Red 2 causes birth defects in animals. At the same time, the FDA announced that all other artificial colors would be tested for their effect on animal fetuses.

Red 2 is the most widely used food coloring, accounting for about one third of all colorings. More than 1,200,-000 pounds were certified for use in fiscal year 1971. The dye is used in soft drinks, ice cream, pistachio nuts, candy, baked goods, pet foods, sausage, breakfast cereals, and other foods.

Although scientists have been concerned since the 1962 thalidomide episode that food additives might cause birth defects, neither the FDA nor manufacturers tested colorings in this regard. The FDA was finally prodded into action in 1971 by a Soviet report, based on amateurish, unreliable experiments, that Red 2 caused birth defects and interfered with rat reproduction. (The Soviet scientists also claimed, again on the basis of poor experiments, that Red 2 caused cancer in rats. However, relatively good experiments conducted on rats, two strains of mice and several dogs supported FDA's contention that Red 2 was not a cancer threat.)

FDA's September statement indicated that severe restrictions would be put on the use of Red 2 by December 31, 1971. But before the new year arrived the FDA asked the National Academy of Sciences to appoint an advisory committee. The Academy agreed to the idea and finally met in early February, but as of March 13 the FDA still

[10] *Ref.: J. Pharm. Exp. Ther. 116* 26 (1956); pers. comm. from FDA.

had not made the committee's recommendations public. Even after the recommendations are released, further delays and legal maneuverings will guarantee the presence of Red 2 in our food for many months, if not years. The Red 2 episode illustrates nicely why many Washingtonians say that FDA stands for Foot Dragging Artists.

There is no reliable method of estimating a chemical's threat to human fetuses on the basis of animal studies. Therefore, if animal studies indicate any chance of birth defects, a chemical as unnecessary as an artificial coloring should be banned without further discussion.

Assuming that Red 2 is eventually restricted or banned, the consumer's troubles will not be over. This dye may be replaced in many foods by a mixture of Red 40 and Violet 1. The hazards of the violet coloring are described above. In addition, we should keep in mind that Red 2 is not the only dye that could cause birth defects; it is simply the only one that has been tested.[11]

Red 3 (Erythrosine). Lifetime feeding studies done by FDA biologists on rats and gerbils indicated that Red 3 does not cause cancer. The thyroid gland is not affected by this iodine-containing dye. Red 3 is used to color cherries in canned fruit cocktail, because it is insoluble in acidic solutions and therefore does not stain other fruit. Biochemists have found that this dye is poorly absorbed and not altered by the rat; its metabolism in man has not been studied. In fiscal year 1970 the FDA certified 154,288 pounds of Red 3.[12]

[11] Ref.: FAO(38B)-22; J. Pharm. Exp. Ther. 136 259 (1962); J. Pharm. Pharmacol. 10 625 (1968); Chem. Abst. 69 1366; Vop. Pitan. 29 (2) 66, (5) 61 (1970); BIBRA (British Industrial Biological Research Association) Information Bulletin, May 1971 (review of Soviet studies); Washington Post, November 14, 1971; February 11, 1972.
[12] Ref.: FAO(38B)-50; FAO(46A)-33; J. Pharm. Exp. Ther. 137 141 (1962); J. Sci. Fd. Agr. 13 650 (1962); Tox. Appl. Pharm. 17 300 (1970).

Red 4 (Ponceau SX). This dye was formerly one of the most widely used, but in 1965 the FDA banned it. Experiments directed by Dr. Kent Davis of the FDA indicated that high levels damaged the adrenal gland and the urinary bladder of dogs. Lifetime feeding studies done at the same time on mice and rats, however, indicated complete safety.

The maraschino cherry industry, which thought that its profits depended upon the use of Red 4, convinced FDA administrators that while large amounts of dye might indeed cause part of the adrenal cortex to atrophy and produce changes in the urinary bladder, the amount of dye that one might ingest by eating artificially colored cherries would have no effect whatsoever on anyone. The FDA now permits Red 4 in maraschino cherries (150 ppm), in certain ingested drugs, and in externally applied cosmetics. FDA inspectors certified more than 23,300 pounds of this dye in fiscal year 1970.[13]

Red 40 (Allura Red AC). This is the food industry's newest food coloring, having been approved by the FDA in April 1971. It was developed to replace Red 4. In a letter to the author, Allied Chemical Corporation "appreciated" the author's interest in their new product but refused to say how well they tested it.[14] Other sources have reported that Red 40 does not appear to cause birth defects.

Yellow 5 (Tartrazine). The only lifetime feeding studies involving this second most widely used coal tar dye were done on rats. FDA pathologists did not detect any harmful effects in this small scale test. Dogs whose diet contained up to 2 percent dye for two years suffered

[13] *Ref.:* FAO(38B)-19; *J. Pharm. Pharmacol. 13* 492 (1961); *I. Pharm. Exp. Ther. 136* 259 (1962); *Tox. Appl. Pharm. 8* 306 (1966); pers. comm. from FDA.
[14] *Ref.: Federal Register* April 10, 1971; pers. comm. from Allied Chemical Corp.

Table II FOOD COLORINGS (other than coal tar dyes)

Substance	CFR*	Source	Color	Special restrictions
dried algal meal	8.301	algae	yellow	in chicken feed to enhance the yellow color of skin and eggs
beta-apo-8'-carotenal	8.302	synthetic (or plants)†	yellow to red	15 milligrams per pound
caramel	8.303	heated carbohydrate	dark brown	—
beta-carotene** (provitamin A)	8.304	synthetic (or plants)†	yellow	—
annatto extract	8.305	seeds from annatto, a tropical tree	yellowish red	—
tagetes (Aztec marigold) meal	8.306	*Tagetes* flower petals	yellow	in chicken feed to enhance the yellow color of skin and eggs
paprika (oleoresin)	8.307 (8.308)	ground dried pod of *Capsicum*, a sweet pepper	reddish orange	—
turmeric (oleoresin)	8.309 (8.310)	ground rhizome of Curcuma, an East Indian herb	yellow	—
saffron	8.311	dried stigma of *Crocus* plant	deep orange	—
fruit juice	8.313			—
vegetable juice	8.314			—

toasted partially defatted cooked cottonseed flour	8.315	cottonseed	brown	—
titanium dioxide	8.316	synthetic	white	up to 1 percent in food
cochineal extract (carmine)	8.317	dried body of cochineal insect	red	—
grape skin extract	8.318	grape	purple red	coloring of beverages
ultramarine blue	8.319	synthetic $Na_7Al_6Si_6O_{24}S_8$	blue	coloring salt for animal feed (up to 0.5 percent)
ferrous gluconate**	8.320	synthetic	develops black color	coloring of ripe (black) olives
dehydrated beets	8.321	beets	dark red	—
corn endosperm oil	8.322	yellow corn grain	reddish brown	in chicken feed to enhance the yellow color of skin and eggs
riboflavin (vitamin B-2)	8.323	synthetic (or natural)†	yellow	—
carrot oil	8.324	carrots	orange	—
iron oxide	8.325	synthetic	reddish brown	dog and cat food (up to 0.25 percent)
canthaxanthin	8.326	synthetic (or natural)†	orange to red	30 milligrams per pound

* section in title 21 in the Code of Federal Regulations.
** see separate entry in this book.
† occurs naturally but the synthetic chemical is more economical.

no ill effects. Lifetime feeding and metabolic studies need to be done in species other than rats. The FDA certified more than 956,600 pounds of Yellow 5 in fiscal year 1970.[15]

Yellow 6 (Sunset Yellow FCF). The third most widely used dye is one of the best studied. Lifetime feeding tests on rats (two strains) and mice (two strains) and shorter tests on rats, mice, and miniature pigs indicated that Yellow 6 is safe. In a seven-year study of dogs, however, the dye (2 percent of the diet) affected the eye, sometimes causing blindness. It is unlikely that the small amount of Yellow 6 in our food (approximately 0.001 percent) would affect our eyes, but to be on the safe side, additional feeding studies should be conducted in other species (for example, the monkey). The metabolism of Yellow 6 in man and dog should also be examined and compared to the way the rat metabolizes it. In fiscal year 1970 FDA inspectors certified 939,641 pounds of this dye.[16]

II. Food Colorings Other Than Coal Tar Dyes

Artificial colorings derived from plants, insects, and minerals were used in food long before coal tar dyes were developed. Currently they account for only about 5 percent of the artificial food coloring used in the United States, because of their technological drawbacks. Table 2 lists the substances currently approved by the FDA for use in foods. Sodium nitrite and nitrate and sodium erythorbate, which contribute to the red or pink color of bacon, ham, hot dogs, and luncheon meat, are not classified by the FDA as artificial colorings; these chemicals are described in separate entries in this book.

[15] Ref.: FAO(38B)-88; Tox. Appl. Pharm. 6 621 (1964); Ann. Allergy 26 309 (1968); Fd. Cos. Tox. 7 287 (1969).
[16] Ref.: FAO(38B)-83; J. Pharm. Exp. Ther. 136 259 (1962); Fd. Cos. Tox. 5 747 (1967), 7 9 (1969); Chiba Daigaku Fuhai Kenkyusho Kokoku 20 101 (1967).

ARTIFICIAL FLAVORING

Flavorings comprise one of the most important classes of food additives, because they are able to replace, or mask the absence of, expensive natural products and to improve the taste of manufactured foods. A measure of their importance and variety is the fact that two thirds of the food additives used in the United States are natural or synthetic flavorings. In 1970 manufacturers spent $150,-000,000 on natural and artificial flavorings, more than any other category of food additive.[17]

If you notice that a food contains "artificial flavoring," you can generally assume that the food contains little or none of the ingredient that would normally supply the flavor. For instance, some brands of strawberry yogurt derive their flavor from real strawberries, while others are based on artificial strawberry flavoring. The peach ice cream that you make at home contains peaches; most commercial peach ice cream contains imitation peach flavoring.[18] Safeway's lemon-flavored gelatin dessert mix contains natural flavoring, while Jell-O brand contains a mixture of artificial and natural flavoring. Substituting artificial flavorings for traditional sources of flavor does permit lower prices or higher profits, but many consumers believe that it also constitutes adulteration.[19] Each person must decide for himself how the difference in taste and in safety between naturally and artificially flavored foods stacks up against the difference in price.

[17] *Chemical and Engineering News*, August 23, 1971, "Food Processing: Search for Growth," Robert M. Hadsell.

[18] "Peach ice cream" contains no artificial flavoring. "Peach flavored ice cream" contains more natural flavoring than artificial flavoring. "Artificially flavored peach ice cream" contains more artificial flavoring than natural flavoring.

[19] According to the law, as long as food is honestly labeled, it is not misbranded or adulterated.

Natural flavors are produced by the combined effect of tens or even hundreds of different chemicals, although frequently the taste of one or two chemicals predominates. Once the basic composition of a natural flavor is determined, scientists can construct a realistic artificial flavoring. The major components may be isolated from a natural source or may be synthesized chemically; in addition, chemicals not present in the natural flavoring may be used.

The job of creating realistic flavors out of purified chemicals is that of taste specialists. These persons, with their highly educated taste buds, prepare and evaluate the tastes of different mixtures and proportions of chemicals. The complexity of their job is reflected in the complexity of flavor recipes. A typical artificial cherry flavoring, for example, is not a single chemical as many persons might assume, but a mixture of thirteen chemicals:

Merory imitation cherry flavor MF 83[20]

1.75	eugenol
4.50	cinnamic aldehyde
6.25	anisyl acetate
9.25	anisic aldehyde
12.50	ethyl oenanthate
15.50	benzyl acetate
25.	vanillin
25.	aldehyde C_{16} (strawberry aldehyde)
37.25	ethyl butyrate
50.	amyl butyrate
125.	tolyl aldehyde
558.	benzaldehyde (primary flavor)
130.	alcohol-95% (solvent)
1000.	

[20] *Food Flavorings*, Merory, J., AVI Publishing Co., Westport, Conn. (1968). (Amount of ingredient expressed in parts per thousand.)

The Food, Drug and Cosmetic Act allows spices and flavorings in a food to be listed on the label under a general term instead of their specific names.

The government regulates flavorings more leniently than most other food additives. There are several reasons for this:

—relatively small amounts are used (usually less than 0.03 percent of a food);

—almost all occur naturally;

—they are represented by a powerful industry lobby. That a chemical is used very sparingly in food, reduces but of course does not eliminate its potential hazard; that a chemical occurs naturally in foods does not guarantee safety. Safrole, for instance, occurs in sassafras and was used in root beer flavoring until 1960 when scientists found that it causes cancer of the liver. A second natural flavoring, oil of calamus, was discovered in 1967 to cause intestinal tumors.

The food additive law grants companies the right to declare that a chemical is "generally recognized as safe (GRAS)" and use it in food without further ado. Flavorings, while not actually on the GRAS list, are handled similarly. Thus, manufacturers can synthesize a new flavoring, proclaim it safe on the basis of minimal or no testing, and add it to our food without obtaining permission from the FDA. Ordinarily, the public is safeguarded from dangerous GRAS compounds, because GRAS compounds, like any other food additives, must be declared on the label. If FDA inspectors discover what they believe to be an unsafe chemical in food, the FDA can challenge the food manufacturer. But this safeguard does not work for flavorings, because they are not identified specifically on the label. A company could call arsenic an "artificial flavoring" and add it to a food without the FDA or the American public ever finding out (except the hard way).

The public's only insurance against dangerous flavorings is that the flavor industry periodically informs the FDA of new substances that they are using in foods.

In the past the industry, through its Washington lobby, the Flavor and Extract Manufacturers' Association, has strenuously opposed more rigorous testing and labeling requirements. Dr. Richard L. Hall, research director of McCormick and Co. (the largest producer of artificial flavorings), and Dr. Bernard L. Oser, president of a large testing laboratory and frequent apologist for the food industry, recently expressed industry's attitude toward untested flavorings. They wrote:

> Scientific progress should not militate against the assumed safety of these substances under their customary conditions of use. To do so would be to penalize research and ingenuity.[21]

In other words, let's continue to assume that these chemicals are safe rather than conduct or believe tests that might indicate otherwise. Such quaint disdain for the public welfare would be laughable were it not coming from the mouths of powerful industry spokesmen.

Most of the thousands of flavorings—natural or synthetic—have not been tested for their capacity to cause cancer, birth defects, or mutations, yet they are consumed daily by virtually the entire American population. The FDA with the assistance of industry and academic scientists should identify the flavorings that are used in the greatest quantity or, as judged from their chemical structure, pose the greatest hazard. Once the priorities are established, the manufacturers of these chemicals—with financial assistance from those segments of the food industry that use artificial flavorings—should be required to conduct detailed toxicologic studies. In this way, over

[21] *Residue Reviews 24* 7 (1968).

a period of years, we should be able to dispel many of the doubts that hang over this group of ill-tested food additives.

Ref.: Seemingly endless lists of artificial and natural flavorings and spices appear in 21 CFR 121.101, 121.1163-4.

Ascorbic Acid   (vitamin C)   GRAS

One difference between man and most other animals is that the human body cannot synthesize ascorbic acid. The only other animals that have the same genetic defect are the primates, guinea pig, bulbul bird, and Indian fruit-eating bat. For these species ascorbic acid is a vitamin (vitamin C). They must obtain this vital nutrient from their diets.

Persons whose diets contain too little vitamin C suffer from scurvy. The symptoms of mild scurvy are muscular weakness, poor teeth, and bleeding gums. Severe and prolonged deficiencies of vitamin C cause fever, degeneration of muscles, hemorrhaging of internal organs, and painful joints. Some of these symptoms are obviously related to ascorbic acid's role in the synthesis of connective tissue (collagen). Biochemists are searching for additional, perhaps more subtle, functions of the vitamin.

Food manufacturers use ascorbic acid, most of which is produced synthetically, in several ways. First and foremost it serves as a vitamin supplement in beverages, potato flakes, and breakfast foods. Vitamin C is so inexpensive, about $1.50 a pound, that fortifying foods with it should not affect their price (one cent's worth can supply thirty-five persons with their daily requirement).[22]

[22] Manufacturers frequently charge excessive prices for vitamin-enriched foods. For example, General Mills adds ½¢ worth of vitamins to a 12-ounce box of "Wheaties" and calls the new product "Total." The public pays 18¢ more for Total than for Wheaties. Vitamin pills are usually far cheaper sources of vitamins than empty foods that have been fortified. Your best bet, though, is to eat good food.

Ascorbic acid is the only food additive that is used both as an antioxidant to prevent the oxidation of food and as an oxidant to promote the oxidation of food. It operates as an antioxidant to increase the shelf life of food in one of two ways. First, it reacts readily with oxygen. As a result of the reaction, the ascorbic acid loses its nutritive value, but the oxygen is prevented from affecting the taste or appearance of the food. The second way ascorbic acid is used as an antioxidant is in conjunction with BHA, BHT, and propyl gallate, three of the most effective "primary" antioxidants. Ascorbic acid regenerates these antioxidants following the chemical changes they undergo when they stop fat from going rancid. The ascorbic acid added to bologna and other meats acts as an antioxidant.

Bakers use ascorbic acid as an oxidant to improve the properties of dough. Flour contains two enzymes that enable ascorbic acid to affect dough. One enzyme reacts ascorbic acid with oxygen, converting it to dehydroascorbic acid. A second enzyme reacts the dehydroascorbic acid with protein in the dough. This second reaction regenerates ascorbic acid and affects the dough in such a way that it kneads more easily and forms a lighter loaf with a finer texture. The ascorbic acid is entirely destroyed in baking and does not contribute to the nutritive value of the bread. Ascorbic acid is used more widely in Europe than in the United States, because of differences in the wheats.

The recommended daily intake of vitamin C for persons over nine years old is 70 to 100 milligrams, or about what is present in six ounces of fresh orange juice. Yet for many years a few doctors and lay people maintained that much larger amounts of vitamin C could do such things as enhance the treatment of chicken pox, speed the healing of wounds, and cure the common cold. In 1970

Professor Linus Pauling threw the weight of his two Nobel Prizes and of a systematic survey of the medical literature behind the theory that large amounts of vitamin C could protect people against the common cold. Much of the medical community pooh-poohed Pauling's argument (often without reading his book or the medical studies), and some "authorities" openly attacked it. However, at least five studies, including two that were inspired by the raging controversy, indicate that a large daily dose of ascorbic acid can clearly reduce the incidence and/or severity of colds. Pauling recommends taking one or two grams a day routinely, and at the first sign of a cold, taking considerably more. The optimum level will vary according to the individual. Persons with chronic diseases, including gout and diabetes, should consult a doctor before experimenting with large doses of vitamin C.

Ref.: FAO(31)-19; *ECT 2* 747; *Ann. N.Y. Acad. Sci. 92* 1 (1961); *J. Vit. 12* 49 (1966); *Baker's Digest 41* 30 (December 1967); *New York Times,* p. 47, January 3, 1971; *Proc. Nat. Acad. Sci. USA 68* 2678 (1971); *Lancet,* p. 1401, June 24, 1972; *Vitamin C and the Common Cold,* rev. ed., Pauling, L., W. H. Freeman, San Francisco (1971).

Ascorbyl Palmitate   GRAS

Ascorbyl palmitate is formed by combining ascorbic acid (vitamin C) with palmitic acid (derived from fat). The *raison d'être* of ascorbyl palmitate is interesting. Ascorbic acid is a good antioxidant but is not soluble in fats. Palmitic acid, on the other hand, is readily soluble in fats and oils. A clever chemist hit upon the idea of chemically combining ascorbic and palmitic acids, thereby creating a fat-soluble antioxidant.

This additive functions as an antioxidant in shortening in the same fashion as ascorbic acid: it readily reacts with oxygen, preventing the latter from reacting with unsaturated fats and causing rancidity. Ascorbyl palmitate is also used as a source of vitamin C in vitamin pills and fortified foods.

Several studies indicate that ascorbyl palmitate is totally metabolized. The ascorbic acid portion of the molecule is available as vitamin C, and the palmitate portion is converted to energy or fat. In the early 1940s Drs. Fitzhugh and Nelson of the FDA conducted a lifetime feeding study on rats and concluded that this compound is safe.

Ref.: FAO(31)-25; FAO(46A)-149; *Arch. Biochem.* 12 375 (1947); *Chem. Abst.* 65 15843h, 68 66889; *Science* 113 628 (1951).

Azodicarbonamide

Azodicarbonamide is one of the newest and most effective dough conditioning agents employed by the baking industry. Dough conditioners such as this have replaced long months of storage as the standard way of aging flour. Natural and chemical aging have identical chemical effects on flour; both produce more manageable dough and lighter, more voluminous loaves of bread.

It seems likely that azodicarbonamide will capture a large share of the flour-treatment market in the coming years. Not only is it as effective as chlorine dioxide—the current leading bleaching/maturing agent—but it is also a powder and easier and cheaper to use than gaseous chlorine dioxide. It is used at concentrations of 2 to 45 parts per million.

Azodicarbonamide does not bleach flour. Therefore, a baker who wishes bleached as well as artificially aged flour must use this chemical in conjunction with a bleach, such as benzoyl peroxide.

When azodicarbonamide reacts with flour, the additive is rapidly and completely converted to a second compound, biurea. In the same chemical reaction the protein (gluten) of flour is oxidized,[23] accounting for the changes in the dough's properties.

Biurea has a reputation for being inert and insoluble. Digestive enzymes do not affect it and most of it is eliminated in the feces. The 10 percent that the body does absorb is excreted in the urine. When scientists fed massive amounts of azodicarbonamide or biurea to rats and dogs for two years, no effect on appearance, survival, growth, organ weight, or pathology was noted. Rat reproduction and lactation were not affected in a study that spanned three generations of rats.

The nutritive value of vitamins and amino acids in bread is not affected by azodicarbonamide or biurea. By all accounts, then, this food additive appears to be quite safe.

Ref.: FAO(40A)-104; *Baker's Digest 37* 69 (1963); *Cereal Chem. 40* 539 (1963); *Tox. Appl. Pharm. 7* 445 (1965); Am. Inst. of Baking Bulletin #127 (February 1967); 21 CFR 121.1085.

Benzoyl Peroxide   GRAS

Benzoyl peroxide has been an important flour bleach since 1917. Because this chemical bleaches but does not "age" flour, bakers use it in conjunction with "aging" agents (azodicarbonamide, iodate, bromate). Fifty parts per million peroxide is sufficient for most flours.

Benzoyl peroxide is a powder that bleaches flour within twenty-four hours of being mixed with it. As the bleach does its work, most of it decomposes to benzoic acid, which remains in the flour after baking. The benzoic acid residue is not hazardous (see "sodium benzoate").

Lifetime feeding experiments have been conducted on rats and mice (200 per dosage level, an unusually large number). No adverse effects were caused by concentra-

---

[23] When flour oxidizes, cysteine residues in protein react with one another.

tions of peroxide up to one thousand times as high as man might encounter. In one experiment the testes of rats atrophied, but this, apparently, was because benzoyl peroxide destroys vitamin E. The rats whose diet consisted almost entirely of treated flour were simply suffering from vitamin deficiency. (The destruction of vitamin E by flour bleaches is a good reason to avoid bread made from bleached flour.) Short-term tests on dogs further supported the safety of benzoyl peroxide.

The effects of flour treated with benzoyl peroxide on the reproduction of animals has not been studied, but benzoic acid, the breakdown product of this additive, does not interfere with reproduction. The possibility that impurities or minor breakdown products of this additive may cause birth defects or mutations should be studied experimentally.

Ref.: FAO(35)-155; *Fd. Cos. Tox.* 2 527 (1964).

Beta Carotene and Vitamin A　　　　　　GRAS

Beta carotene is a yellow pigment that occurs in many fruits and vegetables, as well as in animal fat. Beta carotene is important in a good diet because the body converts it to vitamin A. Food manufacturers add carotene to margarine, nondairy coffee whiteners, shortening, butter, milk, cake mix, dessert toppings, and other products either as artificial coloring, as a nutritional supplement, or both. We can also obtain vitamin A directly from liver or fish oil. While vitamin A (or beta carotene) is necessary for proper health, overdoses may cause birth defects and poisoning.

The average adult should consume 5,000 units of vitamin A or carotene daily; pregnant women need 8,000 units. This vitamin requirement is easily satisfied by beef liver (30,000 units per serving), chard, kale, spinach, and other greens (12,000 units per serving), or by beans, broccoli, carrots, yellow squash, apricots, and sweet potatoes (approximately 5,000 units per serving). Foods to which carotene is added as a nutrient contain 5,000 to 15,000 units per pound of food.

Vitamin A is a vital part of the light-detection mechanism in the retina of the eye and is also necessary for the normal growth of bones and for the health of epithelial tissue. Moderate deficiencies of vitamin A may impair night vision or adaptation to changes in brightness levels, cause dryness and scaliness of the skin, and interfere with proper formation of tooth enamel. Gross and prolonged dietary deficiencies may cause blindness and interfere with reproduction.

Humans can tolerate large excesses of carotene and vitamin A for moderate lengths of time. Thus, while 5,000 units are required daily, 50,000 to 100,000 units daily have been tolerated for ninety days. The chronic daily ingestion of 100,000 units or more—usually from the overzealous use of high potency vitamin capsules—leads to dry, itchy skin, cracking of the lips, and painful areas over bones.

One sure way to get vitamin A poisoning is to eat polar bear liver. Arctic explorers have written vivid accounts of the headaches, nausea, and diarrhea that came in the wake of feasts featuring bear liver. Toxic quantities of vitamin A accumulate in the bear's liver for a reason that is reminiscent of the way DDT accumulates in fish and birds. Unicellular marine organisms synthesize vitamin A and are eaten by plankton, which store the vitamin. The plankton are eaten by big fish which are eaten by polar bears. At each step in the food chain more and more vitamin is stored in the liver. One ounce of polar bear liver contains as much as a half million units of vitamin A; an average-size serving of four ounces would contain two million units.[24]

Large overdoses of vitamin A cause birth defects in animals. Dr. Jane Robens (formerly with the FDA) and other scientists did experiments on pregnant rats, mice, and hamsters and found that dosages that were approximately 50 to 100 times the dietary requirement caused a significant number of deformed fetuses. The sensitivity of the human embryo to the vitamin is not known, but caution would dictate that women carefully regulate their intake of the vitamin during the first three months of pregnancy. They should not consume vitamin preparations that contain 50,000 or more units of vitamin A per capsule.

Ref.: FAO(38B)-38; *Am. J. Dis. Child.* 85 316 (1953); *Tox. Appl. Pharm.* 2 225 (1960), *16* 88 (1970); CUFP-49, 57; CRC-114, 213-4, 217; *Teratology of the Central Nervous System,* Kalter, H., U. of Chicago Press (1968).

Brominated Vegetable Oil

The soft drink industry seems to have nothing but headaches: first exploding bottles, then cyclamates, dentists always at its throat, environmentalists angry about nonreturnable bottles and cans, and most recently, the ingredient brominated vegetable oil (BVO). This additive may be toxic and should not be allowed in food.

Chemically, BVO is vegetable oil (olive, sesame, corn, or cottonseed) whose density has been increased to that of water by being combined with bromine. Flavoring oils are dissolved in BVO, which is then added to carbonated or noncarbonated fruit-flavored drinks. The lighter-than-water oils are dispersed throughout the drink by BVO, without which they would float to the surface and form

---

[24] An ounce of fish liver oil, especially from swordfish, black sea bass, and soupfin shark, contains between 600,000 and thirty million units of vitamin A.

a ring at the neck of the bottle. BVO also makes the soft drink slightly cloudy, giving the illusion of thickness or "body."

Government regulations allow manufacturers to use BVO in soft drinks without listing it on the label. This makes it difficult to know whether BVO is present in a particular drink, except for diet drinks, which are not covered by the regulation. A few of the drinks that contain BVO are Fresca, Orange Fanta, Patio, Mountain Dew, and Orange Crush.

Beverage manufacturers have used BVO for decades and, in the absence of any scientific studies, always assumed that it was safe. In January 1969 Drs. Ian Munro, E. J. Middleton, and H. C. Grice, scientists at Canada's Food and Drug Directorate, published a study that indicated that this additive was not so safe after all. Rats that ate food containing 0.5 or 2.5 percent BVO for eighty days suffered heart, liver, thyroid, testicle, and kidney damage or changes.

Subsequent studies done in England showed that BVO (or fragments of BVO) accumulate in and become permanent residents of animal tissue. Organic bromine residues in human tissues are much higher in countries in which BVO is used than in other countries (e.g. Germany).

No long-term animal studies have been completed, although several are in progress and scheduled for completion in 1973. The effect of this food additive on reproduction has also not been studied. Thus, no one knows whether BVO causes cancer or birth defects. The dosages below which chronic exposure to BVO has no harmful effects on rats (or man) or below which it does not accumulate in body tissues are not yet known.

A year after the Canadian study revealed that BVO was hazardous, FDA administrators removed it from the list of food additives that are "generally recognized as safe." This action meant that manufacturers had to stop using BVO within six months, unless they filed a food additive petition which contained data supporting the substance's safety. When the deadline arrived, manufacturers were still using it in food but had not filed the required safety data. The FDA had sidestepped the law by permitting manufacturers to use BVO during the two or three years it would take to do biological studies. This flouting of the Food, Drug and Cosmetic Act by the very agency that is charged with enforcing it is being challenged in the courts by attorney James Turner and the author.

Before the dangers of BVO were uncovered, beverage manufacturers used it at levels of 300 parts per million. The legal limit in the United States and Canada is now 15 ppm. BVO was banned in Sweden in 1968 and in Great Britain in September 1970.

The FAO/WHO Expert Committee on Food Additives has had an eye on BVO for several years. The Committee wrote in 1966 that:

> The quantity of brominated oils used is likely to be small. Nevertheless, large quantities of beverages may be consumed over long periods of time, and no evidence was available on the possible accumulation of the brominated fatty acids in the body lipids with subsequent release of bromine. The Committee was unable to evaluate brominated oils since evidence from appropriate long-term studies with special reference to possible cumulative effects was not available.

In 1970 the FAO/WHO Committee wrote:

> Although these substances have been used for some years in soft drinks and fruit juices, no evaluation was made at the ninth committee meeting in 1966 because of lack of suitable data. Since then short-term studies in animals have

demonstrated that high doses cause degenerative cardiac lesions. Furthermore, accumulation of lipid and lipid-bound bromine has been demonstrated in adipose tissue and in intracellular fat of various other tissues, both in man and in experimental animals. This evidence suggests that a human epidemiological problem could arise from the uses of BVOs . . . [BVOs] should not be used as a food additive in the absence of evidence indicating their safety.

Brominated vegetable oil is used almost exclusively in snack-type products that are devoid of redeeming nutrients. The public would not suffer one whit if BVO were banned and products containing it were removed from stores until they were reformulated. Manufacturers would suffer little because there are safe alternatives to BVO.

Ref.: Fd. Cos. Tox. 7 25 (1969), 9 1 (1971); FAO(40)-13; *Food Chemical News,* November 9, 1970 (refers to 1970 FAO/ WHO Committee meeting); *Wall Street Journal,* January 25, 1970; *New York Times,* August 29, 1970; pers. comm. from FDA; *Food Product Development,* May 1971, page 90. 21 CFR 121.1234.

Butylated Hydroxyanisole (BHA)   GRAS
Butylated Hydroxytoluene (BHT)

Two of the nastiest-sounding chemicals in our food supply are butylated hydroxyanisole and butylated hydroxytoluene. To minimize any possible consumer unease, foodmakers usually list them on labels as BHA and BHT. You will find one or both of these closely related chemicals in many vegetable oils and in almost every processed food that contains fat or oil. These additives may increase slightly the shelf life of food by preventing polyunsaturated oils from oxidizing and becoming rancid; they may also protect the fat-soluble vitamins (A, D, E). As we shall see, however, BHA and BHT rarely benefit the consumer and should be avoided whenever possible.

BHA has been used in food since 1947, BHT since

1954. Currently, tens of thousands of pounds of both synthetic chemicals are used annually. Manufacturers add BHA and BHT to breakfast cereal packaging (they migrate onto the cereal itself), chewing gum, convenience foods, vegetable oil, shortening, potato flakes, enriched rice, potato chips, candy, and many other oil-containing products. The total concentration of antioxidants (BHA plus BHT plus whatever other ones may be used) ranges from as low as 0.0001 percent in gelatin desserts to 0.1 percent in chewing gum and dry yeast. The usual concentration is about 0.01 percent. According to the FDA an average American diet contains about 0.0004 percent (4 parts per million) of these preservatives. BHT is much cheaper than BHA, but its use is limited because it is less stable at the high temperatures used to pasteurize food.

As a general rule additives should not be allowed in food unless they benefit the consumer, either nutritionally, financially, aesthetically, or by saving him time. BHT and BHA provide good examples of the unnecessary use of synthetic food additives. For instance, some makers of vegetable oil and potato chips add BHA and BHT to their products. Why these companies add the chemicals is not entirely clear. Perhaps they add them out of habit, or because of outmoded manufacturing techniques, or to extend slightly the shelf lives of their products. However, it *is* clear that the additives are superfluous, because many of their competitors do not use them. Jay's potato chips (a midwest brand), Proctor & Gamble's Crisco shortening, Red Star powdered yeast, Wesson vegetable oil and buttery flavor soy oil, Planter's peanut oil and Safeway's corn oil and gelatin desserts do not contain synthetic antioxidants, but are as tasty, wholesome, and economical as competing products which do. Moreover, Eastman Chemical Products, a major producer of antioxidants, states in one of its publications (ZG-157) that "BHA and/or BHT

are not found to provide significant improvement in the
. . . stability of the vegetable oils." The consumer can
greatly reduce his intake of BHA and BHT by purchasing
preservative-free brands of oil, shortening, yeast, and potato chips. The presence of BHA and BHT must be indicated on the package label.

The question of BHT's safety has had a stormy history.
In 1959 Australian biologists reported experiments in
which BHT apparently caused rats' head hair to fall out,
increased the cholesterol levels in the blood, and caused
rats to be born without eyes. These effects were seen
when their food contained as little as 0.1 percent BHT.

The possibility that BHT caused birth defects alarmed
—with good reason—a lot of people, and scientists dove
back into their laboratories to learn more about the dubi-
ous compound. The outcome of this renewed interest was
a barrage of papers published in 1965 that demonstrated
that BHT was quite safe: neither birth defects nor bald-
ing was detected in any of the studies. The reason for the
discrepancy between the 1959 and later studies was never
determined.

Toxicologists have conducted extensive short-term
studies on rats and dogs. The concentrations of BHT and
BHA used in these investigations were as high as 0.5 percent of the diet, or about five hundred times greater than
levels to which humans might be exposed. BHA caused
no adverse effects. High dosages of BHT caused the liver
to enlarge and to develop high levels of enzymes. Pathologists did not observe the liver changes at low dosages
and believe that the changes reflect the liver's efforts to
metabolize the foreign compound.

Scientists have conducted several good studies to determine whether BHA and BHT interfere with animal
reproduction, but they have made little effort to determine
whether or not they cause cancer. The only studies that

could have detected carcinogenicity were done in the
1940s and 1950s and involved small numbers of only
one species (30 to 40 rats per dosage level); moreover,
only a few selected organs were carefully examined for
tumors. Because BHA and BHT are so widely used, their
manufacturers should commence lifetime feeding studies
in several species of animals immediately. Genetic studies
should also be performed.

Toxicologists have determined the way rats and man
metabolize BHA and BHT. Both chemicals are readily
absorbed in the gut, modified in the liver, then excreted.

Biochemists in England recently discovered levels of
BHT as high as 3 ppm in human fatty tissue. The average
level of BHT was six times as high in the fat of Americans (1.3 ppm) as in the fat of Englishmen (0.2 ppm).
Whether the residue of BHT in fat is beneficial or harmful is not yet clear. There is fragmentary evidence that
natural antioxidants, such as vitamin E, retard aging and
increase an animal's life span. Dr. A. L. Tappel, a biochemist at the University of California at Davis, has
suggested that because the average American diet is woe-
fully deficient in vitamin E—containing less than one
fourth of the recommended daily allowance—the few parts
per million of the synthetic antioxidant in our diet may
actually be beneficial. Dr. Tappel quickly adds, though,
that it would be far better to increase our consumption
of vitamin E than to rely upon accidental residues of a
synthetic food additive. We should be suspicious of any
synthetic chemical that accumulates in our body; we
should be doubly suspicious when the chemical has not
been adequately tested for carcinogenic effects.

In summary, butylated hydroxyanisole and butylated
hydroxytoluene have not been adequately tested, accumulate in body fat, and are actually superfluous in many of

the foods in which they are used. They certainly cannot be "generally recognized as safe." Until their safety is established, they should be barred from food; safe alternatives are available. Meanwhile, purchase your food carefully and reduce your consumption of these antioxidants.

Ref.: FAO(31)-41, 45; FAO(40A)-28; CRC-Ch. 5; CUFP-9, 10; *Fd. Cos. Tox. 3* No. 3 (1965); *Adv. Fd. Res. 15* 1 (1966); *Geriatrics 23* 97 (October 1968); *Fd. Cos. Tox. 8* 409 (1970); GRAS (up to 0.02 percent of fat content of a food); 21 CFR 121.1034-5.

Caffeine GRAS

Caffeine, a strong stimulant of the central nervous system, occurs naturally in tea leaves and in coffee, cocoa, and kola beans. Caffeine may cause birth defects and should be avoided by pregnant women.

The discovery that coffee beans contain a substance that wards off sleep has been credited to residents of an Arabian monastery. Shepherds noted that goats that ate coffee beans pranced about all night long. The abbot, learning of this, made from the beans a drink which kept him awake during long, prayerful nights.

Caffeine is used as a food additive only in soft drinks. Kola beans contain caffeine, but much is removed during processing, and so an additional quantity is added to many colas to give them extra "spark," "liveliness," or whatever it is that the latest advertising jingle proclaims. Cola drinks and Dr. Pepper must contain caffeine (up to 0.02 percent), but need not list it on the label. Other carbonated drinks may contain caffeine, but must disclose that fact on the label.

The amount of caffeine in a product varies from brand to brand, but the following average quantities were listed in FDA's "fact sheet" on caffeine: Colas contain 40 to 72 milligrams per twelve-ounce serving. A cup of coffee or tea contains about 90 milligrams of caffeine. According to the Hershey Corporation, one ounce of cocoa contains 50 milligrams of caffeine, a one-ounce bar of milk chocolate 3 milligrams, and one ounce of bitter chocolate contains 25 milligrams. "Stay awake" pills contain approximately 110 milligrams of caffeine. The great fondness Americans have for caffeine-containing beverages and pills probably accounts in part for the booming sales of tranquilizers.

Moderate amounts of caffeine have definite effects on the body. Studies on animals and man have shown that the drug allays drowsiness, enables one to work faster and to think more clearly, stimulates the brain, heart muscle, and kidneys, alters the metabolism of fat, dilates the blood vessels, and causes insulin to be released. The drug increases the production of stomach acid and probably contributes to the number and misery of persons suffering from peptic ulcers.

Caffeine stimulates children more strongly than adults and may cause them to be hyperactive and nervous. Next time you hand your child a bottle of cold, refreshing cola, think for a moment of the agony you may go through a couple of hours later when you try to send him or her to bed.

Persons who drink large amounts of coffee—fifteen to twenty cups a day—may develop "caffeinism." The symptoms of this illness, which is most often observed in waitresses, long distance truck drivers, and night-shift workers, are insomnia, a slight fever, and irritability. The nerve-racking symptoms quickly recede when sufferers cut down on their coffee drinking.

For a time heart specialists conjectured that coffee and tea might promote heart disease, which is associated with high fat levels in blood, because caffeine raises fat or cholesterol levels in the blood of laboratory animals. Several ambitious studies were undertaken in the 1950s and 1960s in the United States, Canada, and England to look for a relationship between coffee consumption and heart attacks. Researchers did not detect a significant correlation in any of the experiments.

Caffeine belongs to the class of chemicals called purines, some of which are important constituents of chromosomes and genes. Geneticists knew that some purines caused mutations and feared that caffeine, a widely consumed chemical, might also cause mutations. In the past two decades scientists have conducted a variety of experiments to evaluate the potential hazard. The earliest studies, which were done around 1950, showed that caffeine was weakly mutagenic in bacteria. This positive finding on one of the lowest forms of life encouraged scientists to do experiments on mammals and insects. In 1968 a German researcher claimed to have shown that caffeine caused mutations in mice. However, other scientists found flaws both in his data and in his conclusions. Subsequent careful studies by eminent geneticists and toxicologists all gave unequivocally negative results.[25] The consensus of scientists now is that caffeine does not cause mutations in humans.

While caffeine does not cause mutations, it does cause birth defects, at least in animals. Dr. H. Nishimura and his colleagues at Kyoto University found that injecting 100–200 milligrams of caffeine per kilogram of body weight into pregnant mice induced birth defects in 6 to 20 percent of the offspring. These dosages were quite high, equivalent to injecting into a woman the caffeine contained in 50 to 100 cups of coffee. In three additional studies conducted in Germany, France, and England, caf-

[25] Dominant lethal mutations, recessive mutations, and damaged meiotic chromosomes were looked for and not found.

feine was fed—not injected—to pregnant mice in amounts corresponding to 25 cups of coffee per day for a woman (50 to 75 milligrams per kilogram). Birth defects occurred in 1 to 3 percent of the baby mice in two of the studies but were not observed in the third. Higher oral dosages of caffeine, 100 to 150 milligrams per kilogram, caused malformations in 8 to 20 percent of the fetuses, respectively. The proportionality between dose and incidence of malformations increases the confidence we may have in these experiments and indicates that even the small amount of caffeine present in a cup or two of coffee a day might cause birth defects.

Experiments on animals provide only a rough measure of a chemical's effects on human fetuses. Humans are sometimes much more sensitive (thalidomide) or much less sensitive than other species of animals to chemicals that cause birth defects. So one cannot state conclusively that caffeine does or does not cause birth defects in humans. Clearly, we need more information: we need surveys of the caffeine intakes of women who had miscarriages or who gave birth to malformed babies, and we need further teratology studies involving different species of animals and lower levels of caffeine. However, the animal experiments point to the existence of a real hazard. Obstetricians and books about childbearing should inform expectant mothers of the possible danger.[26] *As a reasonable precaution, women in the first three months of pregnancy should reduce or totally eliminate their consumption of caffeine-containing beverages, foods, and*

----

[26] Coffee causes nausea in many pregnant women; the body may have evolved this mechanism for discouraging women from consuming a possibly dangerous chemical.

Mrs. Sandy Ward and the author surveyed a score of obstetricians in the Washington area and discovered that some advised patients not to use caffeine stay-awake tablets, but all told patients that it was all right to drink coffee.

drugs. For inveterate coffee-drinkers a decaffeinated coffee such as Sanka provides an alternative.

Ref.: Arzneimittel-Forschung 14 415 (1964); *C.R. Soc. Biol. 159* 2199 (1965); *Lancet 1* 721 (1966); *Postgrad. Med. 44* 196 (1968); *Arch. Biochem. 134* 434 (1969); *Jap. J. Pharm. 19* 134 (1969); *Metabolism 18* 1007 (1969); *Fd. Cos. Tox. 8* 381 (1970); pers. comm. from Hershey Foods; GRAS (up to 0.02 percent in soft drinks).

Calcium and Sodium Propionate GRAS

Almost everyone has heard of the food additive calcium propionate. It is the chemical most often used to prevent the growth of mold and certain bacteria in bread and rolls. While preserving the baked goods, it also provides a calcium supplement to the diet. Sodium propionate, the sodium salt of propionic acid, is preferred in pies and cakes, because calcium alters the action of chemical leavening agents.

Propionic acid occurs naturally in many foods and acts as a natural preservative in Swiss cheese, in which it may occur at a 1 percent level. Propionate is also formed and used as a source of energy when the body metabolizes certain fats and amino acids.

Because most germs are killed when cakes and breads are baked, one might wonder why preservatives are needed at all in these foods. The most obvious reason is that molds may contaminate a food after it is removed from the oven. Secondly, during baking the interior of a loaf of bread may never rise above 100° C (212° F), which is not hot enough to kill all bacterial spores. Were it not for a preservative these bacteria would multiply and soften and discolor the bread and cause off-odors.

Propionates inhibit the growth of, but do not kill, germs. Although germs tolerate small amounts of propionate, the moderate levels (0.1 to 0.2 percent) of propionate used in food appear enormous from the humble perspective of a germ. When certain chemicals in molds and bacteria are confronted with the onslaught of propionate, they are excessively occupied with one metabolic function and are not available to perform other vital functions. Consequently, the germs grow slowly or not at all. The yeast used to make bread rise is added at levels high enough to resist the effects of the preservative; the yeast is killed during baking.

Propionate is probably one of the most innocuous food additives. Fed to rats as 3.75 percent of their diet for a year, sodium propionate had no ill effects. A 15 percent solution of propionate applied to a man's eye caused no irritation.

Ref.: FAO(31)-84; CRC-249; *Chem. Abst. 69* 33733.

Calcium and Sodium Stearoyl Lactylates and Sodium Stearoyl Fumarate

Stearoyl lactylates and stearoyl fumarate are closely related chemicals that were first used commercially in the early 1960s. These substances strengthen bread dough so that it can withstand the mechanical punishment it receives in modern bread-making machinery; moreover, bakers do not have to control mixing times as precisely. Bread and cake made with these additives (plus mono- and diglycerides) stay soft longer and have a more uniform grain and greater volume than untreated products.

Manufacturers use stearoyl lactylates and fumarate in bread dough, cake fillings, and other starch-containing foods at levels up to 0.5 percent. These chemicals also serve as whipping agents in dried, liquid, and frozen egg whites and in artificial whipped cream.

Biochemists have found that digestive enzymes in animals break these additives down to stearic alcohol and either lactic or fumaric acid, all of which are harmless,

easily metabolized substances. No lifetime feeding or reproduction studies have been conducted.

Ref.: FAO(40)-15; FAO(46A)-109; C&EN-123; CUFP-26;
21 CFR 121.1047-1048, 121.1183.

Carrageenan   GRAS

For centuries, housewives living near the coasts of Ireland and France have used a substance obtained from
the sun-bleached fronds of a local seaweed, Irish moss,
as an ingredient in blancmange pudding. Europeans have
also used the same seaweed to treat peptic and intestinal
ulcers. The active chemical in Irish moss is the carbohydrate carrageenan, named after Carragheen, Ireland. As
discussed below, infant formulas that contain carrageenan
may not be safe.

Irish moss grows along the shores of the Maritime
provinces of Canada, Maine, the British Isles, Scandi-
navia, and France. Fishermen scrape the red, bushy seaweed from coastal rocks with special long-handled rakes,
or simply collect it on the beach. Carrageenan is extracted
from the plant with hot water.

Food manufacturers use carrageenan because of the
unique way it reacts with milk protein. A weak gel that
forms in the milk prevents cocoa particles from settling
in chocolate milk and prevents butterfat from separating
out of evaporated milk. The seaweed derivative has similar stabilizing effects in canned milk drinks, "instant
breakfast," and infant formula. Manufacturers use carrageenan to add "body" to soft drinks, to thicken ice
cream, jelly, sour cream, and syrup, to stabilize the foam
in beer, and to prevent the oil from separating out of
frozen whipped topping.

Carrageenan is also gaining favor as the gelling component of gelatin-type desserts and milk puddings, although its rising cost may limit its use. A carrageenan gel

forms more readily and melts less readily than a gel made of gelatin, and it does not develop a thick skin on standing. Carrageenan gels have a natural brittleness which may be tempered by adding a small amount of locust bean gum. One disadvantage of a carrageenan gel compared to one made from gelatin (a protein) is that it has no nutritional value. Carrageenan-based products are used by vegetarians, because gelatin is made from animals.

Scientists have fed carrageenan to animals and observed that it is poorly absorbed by the body and is apparently harmless at levels as high as 15 percent of the diet. More recently, however, British scientists discovered that the guinea pig and rabbit may be sensitive to carrageenan. When guinea pigs drank water containing 1 percent carrageenan for one month, they developed ulcers in the caecum (the pouch at the junction of the large and small intestines) and, to a lesser extent, in the large intestine. When carrageenan was treated with acid, which decreases the size of the molecules, and fed to rabbits and guinea pigs, levels as low as 0.1 percent of the diet caused intestinal ulcers.

Companies that manufacture canned infant formula use carrageenan to prolong the shelf life of their products. Whether the substance that causes ulcers in guinea pigs also jeopardizes the health of babies is not yet known. The hazard depends upon the amount of carrageenan used and the size of the molecules. The author's inquiries about these points to the Research Directors of three manufacturers of formula (Mead Johnson—"Enfamil"; Wyeth Laboratories—"SMA"; Ross Laboratories—"Similac") were never answered.

One factor that reduces the hazard to babies is that they always consume carrageenan with milk, which reduces the acidity of the additive; in the animal studies carrageenan was added to the drinking water. That the sensitive laboratory animals were herbivorous, may also be significant. On the other hand, the formula "Similac" contains polysorbate 80 along with carrageenan; the combined effect of the two food additives has never been tested.

As far as could be determined, scientists have not studied the effect of carrageenan on infants. Mead Johnson, Wyeth, and Ross refused to say whether they have conducted such studies.

Doctors James Watt and R. Marcus, the British pathologists who discovered the ulcerative effect of carrageenan in animals, believe that this food additive may be hazardous. In a 1970 scientific paper, they wrote:

> In the preparation or digestion of carrageenan-containing foods, some degree of degradation probably occurs. Although the cause of ulcerative colitis in man . . . is unknown, it is possible that carrageenan may sometimes be an important etiological factor . . . it is evident that the use of carrageenan either as a food additive or as a drug will have to be reconsidered.

The wisdom of exposing infants to carrageenan is highly questionable and deserves a thorough review by the Food and Drug Administration. Meanwhile, this additive should be removed from infant formulas and replaced, if a replacement is indeed necessary, with a safer substance. Inasmuch as British infant formula does not contain carrageenan, this additive is hardly irreplaceable.

Ref.: FAO(46A)-93; CRC-325-7, 348; *Wall Street Journal,* September 24, 1970; *J. Pharm. Pharmacol 21* 187S (1969); *Gastroenterology 59* 760 (1970); pers. comm. from Unigate Foods, Ltd. England).

Casein   GRAS

A science experiment that millions of boys and girls have conducted at their breakfast tables is to add grapefruit juice or lime juice to milk. Like in a magic trick, mixing the two liquids together causes the milk to curdle. The curd is primarily casein, the principal protein in milk. Casein is a nutritious protein, because it contains adequate amounts of all the essential amino acids. Food manufacturers add casein to ice cream, ice milk, frozen custard, and sherbet to improve their texture; in nondairy coffee creamers, casein adds body and acts as a whitener.

Chewing Gum Base

Chewing gum base is the tasteless wad that is left in your mouth when all the flavor is gone from a stick of gum. Gum base is composed of natural (chicle, natural rubber, etc.) or synthetic (butyl rubber, paraffin, polyethylene, etc.) masticatory substances, synthetic softeners (glycerated gum resin, stearic acid, etc.), resins, antioxidants, and other goodies. As you might expect, chewing gum bases have no nutritive value, and they have not been adequately tested.

Turn to the discussions of mannitol and sorbitol for a word about noncariogenic ("sugarless") chewing gum.

Ref.: 21 CFR 121.1059 lists some of the dozens of substances that gum manufacturers use.

Chlorine (Cl₂)

Before bakers can use flour, they must "age" it, either by storing it for several months or by treating it with a chemical. The best agent for aging cake flour is chlorine gas, which, at a level of 400 parts per million, instantly ages the flour and bleaches it white.

Chlorine, like other aging agents, affects the flour by causing changes in the protein fraction of flour. In addition, chlorine reacts to a limited extent with the small amount of unsaturated oil that is present in white flour

Toxicologists were suspicious of the chlorinated flour oil (CFO) that forms and therefore conducted tests on rodents to determine its possible effects.

In one series of experiments, British scientists treated flour with chlorine and then extracted the CFO. They fed the modified oil at high concentrations (up to a thousand times the amount to which a person might be exposed) to four generations of rats and mice. CFO did not accumulate in body fat and had an adverse effect (on lactation) only at the highest concentration. While CFO appeared to be innocuous, it should be noted that very few animals (2 to 6 mother rodents per dosage level per generation) were used in the experiment.

The FAO/WHO Expert Committee on Food Additives recommended that further studies be conducted: "Adequate long-term studies are needed on flour treated with chlorine at several dose levels and [on] bread baked from it."

Ref.: FAO(40A)-109; Proc. Nutr. Soc. 25 51 (1966).

Chlorine Dioxide (ClO$_2$)   GRAS

In 1946 the baking industries in the United States and abroad were rudely shocked. Sir Edward Mellanby, an English biochemist, demonstrated that ordinary flour or bread that had been treated with agene (nitrogen trichloride) made dogs stark raving mad and was sometimes even fatal. At the time, agene was used to bleach and age 30 to 90 percent of the bread flour in the United States and England. Cereal chemists anxiously sought an economical, effective, and safe substitute and found that one of the lesser used flour treatment agents, chlorine dioxide, filled the bill. After agene was banned in 1949, chlorine dioxide gas became the most widely used bleaching and maturing agent.

Chlorine dioxide works by releasing oxygen which causes changes in the protein fraction of the flour and which bleaches the yellow pigments. Fourteen parts per million of chlorine dioxide has an immediate bleaching and maturing effect and is the normal amount used to treat flour.

Chlorine dioxide has been rather well tested, because no one wanted a repeat of the agene scare. In one of the more complete experiments, four generations of rats were fed diets consisting mainly of heavily treated (300 ppm) flour or bread. No adverse effects on growth, survival, internal organs, or reproduction were observed. Short-term tests have been done in the rat, rabbit, dog, and monkey.

Chlorine dioxide and other bleaching agents reduce the vitamin E content of flour. Whether this loss has any nutritional significance, though, is doubtful, because even unbleached white flour contains little of the vitamin.[27] Most Americans consume inadequate amounts of vitamin E; what little we do consume comes mainly from vegetable oil. See "alpha tocopherol."

Ref.: FAO(35)-159; *J. Sci. Fd. Agr.* 7 464 (1956), *15* 725 (1964), *18* 203 (1967).

Citric Acid   (sodium citrate)   GRAS

Citric acid is one of the most versatile, commonly used, and unquestionably safe food additives. It is an important metabolite in virtually all living organisms and occurs in high concentrations in citrus fruits and berries. It comprises as much as 8 percent of the juice of unripe lemons and a somewhat smaller proportion of the juices of many ripe fruits.

---

[27] Most of the vitamin E in a grain of wheat resides in the germ, which is eliminated during the milling and refining operations. The greater vitamin E content is one of the main reasons why whole wheat bread and flour are more nutritious than white bread and flour.

Chemical companies have been producing citric acid since 1860. They normally make it by the fermentative action of fungus (*Aspergillus niger*) on beet molasses. The growing conditions are carefully controlled so that the fungus produces but does not metabolize the acid.[28] A small amount is also isolated from pineapple by-products and from low-quality lemons. American food companies use fifty million pounds of citric acid annually.

Citric acid is an important food additive because it is a strong acid, inexpensive, has a tart flavor, and serves as an antioxidant. The food industry uses it in ice cream, sherbet, fruit juice drinks, carbonated beverages, jellies, preserves, canned fruits and vegetables, cheeses, candies, and chewing gum. In soft-centered candy citric acid has the added function of solubilizing the sugar (see "invert sugar").

Manufacturers use citric acid as an antioxidant in instant potatoes, wheat chips, and potato sticks. Citric acid prevents spoilage by trapping metal ions which might otherwise promote reactions that spoil or discolor the food. It is often used together with "primary" antioxidants (BHA, BHT, propyl gallate), which inhibit oxidation by a totally different mechanism. The combined effect of the different kinds of antioxidants is synergistic. Wine producers make good use of citric acid's ability to trap metal ions; the acid combines with iron, which, if free, forms iron-tannin complexes that make wine cloudy.

You will rarely find citric acid used as an acid in powdered products, such as gelatin desserts or powdered beverage mixes, because it tends to attract moisture and cause caking.

---

[28] Copper is added to inhibit aconitase, the enzyme that metabolizes citrate; metal atoms that are components of enzymes that promote the degradation of citrate are preferentially removed; the acidity of the growth medium is kept below pH 3.5.

The sodium salt of citric acid is used as a buffering agent to control the acidity of gelatin desserts, jams, jellies, candies, ice cream, and other products.

Ref.: FAO(31)-50; FAO(35)-54; *ECT 5* 524; CRC-Ch. 7.

Corn Syrup   GRAS

Corn syrup is a sweet, thick solution made by digesting cornstarch with acids or enzymes. It contains dextrose, maltose, and dextrin. Food manufacturers use corn syrup to sweeten and thicken foods and beverages, but in some foods it also serves additional purposes. It retards crystallization of sugar in candy, icings, and fillings, and prevents the loss of moisture from cakes, cookies, and whipped foods.

Corn syrup may be dried and used in powdered products, such as nondairy coffee whiteners, in which case it is called corn syrup solids.

Ref.: ECT 9 926.

Cyclamates

"Health Hazard" . . . "Bladder Cancer" . . . "Mass Human Experiment" . . . "Severe Blow to Canners" . . . "Cyclamate Nightmare" . . . These were some of the headlines and phrases making the news in 1969–70 when the story broke that cyclamate, a widely used sugar substitute, could cause cancer. In what was certainly the most hectic period in recent years for the Food and Drug Administration, the public got rare insight into the workings of the government agency that is charged with keeping our foods free of toxic chemicals. That agency's procrastinations and prevarications surfaced time and time again, in a seemingly endless stream. Two years later, the waves generated by a decision affecting $1 billion worth of food have still not subsided.

Cyclamates were originally added to food to provide a low-calorie, nonsugar sweetener for the benefit of both dieters and diabetics. Around 1960, beverage and other manufacturers began pushing "diet" foods to the general public. Industry's advertising pitch was that if you bought diet foods, you would lose weight. Sales of low-calorie drinks and foods soared as people were persuaded to be increasingly concerned about their waistlines. The new products opened up a whole new market, and because cyclamates could do as good a sweetening job as sugar at a lower cost, the profits were enormous.

As the use of cyclamate increased, doctors and scientists began taking a closer look at its effectiveness in helping people lose weight and at its safety. Some nutritionists alleged that the chemical really did help people lose weight; other scientists produced studies indicating that it was worthless. The general consensus appeared to be that you would not lose weight just by buying a bottle of Fresca instead of Coke; to lose weight you had to make a strong, total-diet effort to cut down on calories. On the safety issue, the National Academy of Sciences warned in 1954, 1955, 1962, and 1968 that the artificial sweeteners should not be used by the general public. All the warnings went unheeded by industry, the government, and the public, and in 1969 the storm broke: evidence began pouring in to the FDA that cyclamate caused bladder cancer, birth defects, and mutations in animals. Because cancer takes ten to twenty years to develop in humans, we do not know yet whether we shall be witnessing a cancer epidemic among those who ate diet foods.

Once there was good evidence that cyclamate was carcinogenic and teratogenic, the FDA did not comply with the law that prohibits the use in food of unsafe chemicals, particularly those that cause cancer (the so-called Delaney Amendment). Instead of banning cyclamate, the FDA recommended in April 1969 that people restrict their intake of it. After this equivocation, public pressure and evidence that cyclamate caused cancer both mounted steadily. In October the FDA banned cyclamate totally and ruled that no cyclamate-containing products could be sold after February 1, 1970. In the next few months, however, the food industry gained the upper hand, and in February the FDA reversed itself. This time the FDA Commissioner said that cyclamate could be used in food, if the foods were labeled "drugs"; manufacturers had until September 1, 1970, to switch their labels. Finally, on August 14, 1970, FDA administrators succumbed to public, scientific, and congressional pressures, reversed directions again, and banned cyclamate totally and forever; stores were given two weeks to get the affected products off their shelves.

As FDA's machinations came to an end, industry developed a two-pronged strategy to recoup some of its multimillion-dollar losses. First they tried dumping their products for whatever they could get. If they could sell them for a few dollars, fine, otherwise they donated them to charity to reap substantial tax savings. Carnation, for instance, saved $2 million in taxes. After September 1, 1970, you could find the banned products wherever there were poor people. Cases of soda were piled high in "thrift" stores across the United States, and on a ship in the Pacific, sixty thousand cases of Carnation's "Slender" low-calorie drink were headed for calorie-starved Laotian refugees. The shipment to Laos, at a cost of $42,000 in taxpayer dollars for shipping fees, was conceived as a way of getting a tax break. Fortunately, Ohio Congressman Charles Vanik, calling Carnation's plan "cheap and cruel," prevented distribution of the carcinogen-containing product.

Industry's second tactic was to get the U. S. Treasury

to reimburse firms for their losses. In September 1971 the House Judiciary Committee heard testimony on legislation that would do just that. As of March 1972 the Nixon-supported bill is progressing through Congress.

The cyclamate episode gave citizens a chance to observe the way scientists think and act during a health crisis. As one would expect, there were some scientific heroes and some scientific goats. Several brave FDA scientists, including Dr. Jacqueline Verrett and Dr. Marvin Legator, brought the cyclamate matter to the public's attention after their views had been squelched deep within the bureaucratic bowels of the FDA. Pressure from these and other scientists helped force the FDA to act. Meanwhile, other scientists were saying that ingesting a tiny amount of a carcinogen is OK, or that cyclamate may cause cancer in animals but we don't know for sure if it causes cancer in humans, so it should not be banned. The latter view was audaciously advanced by Dr. Melvin Benarde, a professor at the Hahnemann Medical College in Philadelphia. Benarde wrote:

> Conversely, if an additive is found to be carcinogenic in animals, does it mean it will do the same in humans? *Unfortunately*, we can never learn this because the Delaney Amendment of 1958 proscribes the use of such an additive for man once it is found hazardous to animals (italics added).[29]

Contrast Benarde's philosophy with that of Dr. Legator and five other scientists who are concerned about our chemical environment:

> The decision to restrict cyclamates to the general public and to terminate a mass human experiment for which there are no demonstrable matching benefits is clearly proper.

[29] Benarde, Melvin A., *The Chemicals We Eat*, American Heritage Press, New York (1971), pp. 126–27.

We concur that food additives be banned from products unless they have been proven safe, and either significantly improve the quality or nutritive value of the food or lower the food cost.[30]

Looking back at the cyclamate brouhaha from the perspective of two years, it appears that the public has benefited in three ways from what transpired. First, cyclamate has been banned. Second, the pressure and scrutiny to which FDA administrators were subjected has encouraged them to act more responsibly. Third, everyone—the public, scientists, and FDA staff—is approaching food additives with a much more skeptical and conservative eye.

Ref.: Wall Street Journal, August 7, 1970; *Washington Post,* August 15, 1970; Turner, J., *The Chemical Feast,* Grossman, New York (1970); "Report on Nonnutritive Sweeteners," FDA (October 16, 1971); "Cyclamate Report," Appendix I, FDA (July 1970).

Cysteine   GRAS

Cysteine is an amino acid and therefore a natural constituent of protein-containing foods. Manufacturers add small amounts of cysteine to foods to prevent oxygen from destroying vitamin C. Bakers use cysteine to enhance the dough-improving effects of flour treatment agents.

Ref.: Baker's Digest 41 (3) 34 (1967).

Dextrin   GRAS

The mixture of fragments that results from treating starch with acid, alkali, or enzymes is called dextrin. Dextrin is as safe and nutritious as starch itself. See the entry for starch.

Food manufacturers use dextrin to prevent sugar from

[30] Epstein, S. S., Hollaender, A., Lederberg, J., Legator, M., Richardson, H., Wolff, A. H., *Science 166* 1575 (1969).

crystallizing in candy, to encapsulate flavor oils used in powdered mixes, and as a thickening agent.

Dextrose   (glucose)   GRAS

Dextrose is one of the most important and ubiquitous chemicals in living organisms. This chemical is present in every living cell. Plants obtain dextrose from photosynthesis, while animals make dextrose from a variety of different chemicals or obtain it from food. Dextrose has a key position in the metabolic pathways of both plants and animals; when an organism converts chemical "A" to chemical "Z", it often converts "A" to dextrose, and then dextrose to "Z". Dextrose is a sugar and the source of sweetness in fruits and honey.

In 1811 a Russian chemist named Kirchoff discovered that he could produce dextrose simply by heating starch in the presence of acid. For this valuable discovery the Russian Tsar awarded Kirchoff with a lifetime pension of five hundred rubles a year and the Order of St. Anne. More recently, organic chemists have found that starch, cellulose, and glycogen (animal starch) consist of great numbers of dextrose molecules linked end to end. When we eat plant or animal starch our digestive enzymes break it down to dextrose. Our body uses the dextrose for energy or converts it to glycogen, which is stored in our liver and muscles. When we need energy our body converts glycogen back to dextrose. Man and most other animals cannot digest cellulose, because we lack certain enzymes.

Food manufacturers use dextrose primarily as a sweetener, but it serves additional functions in certain foods. Dextrose (and other sugars) turns brown when heated and contributes to the color of bread crust and toast. The same browning reaction accounts for the brown color of caramel. In soft drinks, dextrose contributes "body" and "mouthfeel." Dextrose is only three fourths as sweet as table sugar (sucrose) and may be used in place of table sugar when oversweetness is a problem.

Ref.: ECT 6 919.

Diethyl Pyrocarbonate

The FDA's final food additive headache in 1971 derived from diethyl pyrocarbonate (DEPC), a chemical that prevents microorganisms from growing in alcoholic beverages and fruit drinks. From the time it was introduced in the early 1960s, DEPC was thought to be the ideal beverage preservative, because it quickly kills microbes and then breaks down almost completely within twenty-four hours to carbon dioxide and alcohol, both of which are harmless. This preservative was developed in Germany and used more widely in Europe than in the United States.

DEPC's tragic flaw is that it is extremely reactive and can potentially combine with ammonia to form urethan, a strong carcinogen. For several years analytical chemists unsuccessfully searched for urethan in products treated with DEPC. In December 1971, however, two Swedish researchers used a sensitive technique involving the addition of radioactively labeled DEPC to orange juice, wine, and beer. They found that small but significant amounts of urethan formed in each of the drinks. FDA scientists are attempting to repeat and confirm the Swedish study. Meanwhile, the FDA proposed on February 11, 1972, that this once highly touted preservative be banned.

DEPC is used by a few American producers of draft beer, wine, and fruit drinks, but it is never listed on a food label because by the time the beverages reach the consumer all the DEPC has broken down. Even if it did not break down, however, DEPC would still not be listed on cans and bottles of alcoholic beverages. The reason for this is that the Internal Revenue Service regulates the labeling of alcoholic beverages[31] and does not require either major ingredients or chemical additives to be listed on the label. The only exception is artificial coloring in distilled spirits. Wine and beer producers can—and do—add any of a long list of chemicals to their products without informing the consumer. A beer may be colored with caramel, sweetened with sugar, treated with propylene glycol alginate to stabilize the head, spiked with EDTA to prevent "gushing" when a can is opened, and preserved with heptyl paraben.

Wine producers are "restricted" to approximately seventy additives. These substances include sulfur dioxide to prevent overfermentation, atmos 300 and antifoam AF to inhibit foaming, and copper sulfate and polyvinylpolypyrrolidone to clarify the wine.

DEPC was not the first chemical added to alcoholic beverages that turned out to be hazardous. In 1966 cobalt was banned as a foam stabilizer in beer after it was discovered to have killed about fifty heavy drinkers in the U.S. and Canada. Experiences such as these suggest strongly that the ingredients in alcoholic beverages should be labeled just as in other foods and drinks. The public has as much right to know what is in its beer as in its crackers. Full labeling would enable consumers to avoid particular additives if they so wished and also to avoid highly doctored brands. Moreover, if an additive is found to be dangerous the government, retailers, and the public could quickly locate products containing that chemical.

Ref. (DEPC): *Science 174* 1248 (1971); Washington *Star,* December 22, 1971; *Federal Register,* February 11, 1972, p. 3060.

Ref. (alcoholic beverages): 26 CFR 240.1051; 27 CFR sections 4, 5, 7.

[31] The FDA still judges the safety of additives used in alcoholic beverages.

Dimethylpolysiloxane　　(methyl silicone, methyl polysilicone)

Silicon, which is present in sand and rock, is the second most abundant element in the earth's crust. Compounds containing silicon, despite their ready availability, are not utilized by living organisms and are not necessary for their growth (except for diatoms, which make their delicate shells from silica).

Silicon is chemically closely related to carbon and, like carbon, can form the backbone of an infinite number of compounds. In the last thirty years chemists have synthesized—and engineers have developed uses for—an incredible variety of silicon compounds, many with highly unusual properties. The silicones, which are long polymers containing silicon, oxygen, and organic groups, are the most important of the synthetic silicon compounds.

The assortment of jobs to which silicones have been applied boggles the mind. Silicones are used in bouncing putty, ironing spray, water-repellent coatings, and electrical insulation; silicone fluids are used to supplement soft tissues of the body (witness some California go-go dancers[32]) and to lubricate arthritic joints; silicone rubber tubing is used to replace damaged tissues in the body, such as the urethra. Most of these applications depend upon silicone's properties of being insoluble in water and chemically inert.

The use of silicone by food manufacturers is trivial compared to its other uses. Dimethylpolysiloxane, the

[32] According to the American Medical Association, "The injection of silicone fluid to increase the size of the female breast is an unapproved surgical technique and dangerous." The FDA has reported that at least four women have died because of silicone entering the bloodstream and becoming lodged in the brain or lungs.

basic silicone, prevents the formation of troublesome foam during the manufacture of wine, refined sugar, yeast, gelatin, and chewing gum. Vegetable oil makers add it to their products, under the name methyl silicone or methyl polysilicone, to prevent foaming and spattering when the oils are cooked. This synthetic compound is used at levels between 0.9 and 10 parts per million.

The safety of silicones, which would be expected on the basis of their inertness, has been supported by feeding tests in animals (up to 0.1 percent of the diet) and by clinical usage in man. In a chronic feeding study done on rats, 0.1 percent silicone had no visible effect on growth, blood composition, tissues, and reproduction. Biochemical studies, designed to indicate the extent to which dimethylpolysiloxane is absorbed by the body, could not be found.

Ref.: FAO(46A)-151; *AMA Arch. Ind. Health 21* 514 (1960); *Chemistry and Technology of Silicones,* Noll, W., Academic Press, New York (1968); CUFP-21; 21 CFR 121.1099.

Dioctyl Sodium Sulfosuccinate (DSS)

A recurring problem that manufacturers have in the factory and cooks have in the kitchen is getting certain powdered foods to dissolve in water. Starch, dry milk, and other items sometimes just refuse to get wet! The difficulty has to do with the surface tension of water, and the solution to the problem is to coat the powder with a very small amount of a detergent-like chemical, such as dioctyl sodium sulfosuccinate (DSS).

The most likely place in the pantry to encounter DSS is in powdered soft drink mixes, in which it helps fumaric acid dissolve in water. Manufacturers also use it in some canned milk beverages containing cocoa fat, and in foods that contain hard-to-dissolve thickening agents. DSS has a wide variety of industrial applications, such as in sugar refining, cleaning fruits and vegetables, and washing bottles.

In the 1940s scientists at American Cyanamid Company and at the FDA conducted one small-scale lifetime feeding test (twelve rats per dosage level) and several shorter tests (on rats, dogs, rabbits, and monkeys). All the tests indicated that dioctyl sodium sulfosuccinate is safe at the levels used in food. American Cyanamid researchers recently studied the effects of DSS on the reproduction of rats. Large amounts of the chemical had no apparent effect on the rats.

Ref.: *J. Am. Pharm. Asso. 37* 29 (1948); *J. Ind. Hyg. Tox. 25* 175 (1943); CRC-256, 498, 622; pers. comm. from American Cyanamid Co., Pearl River, N.Y., regarding testing data; 21 CFR 121.1137.

Disodium Guanylate (GMP)
Disodium Inosinate (IMP)

Disodium guanylate (GMP) and disodium inosinate (IMP) belong to the same family of food additives as monosodium glutamate, the flavor enhancer. Flavor enhancers have little or no taste of their own, but accentuate the natural flavor of foods. They are often used by manufacturers in place of more expensive natural ingredients.[33]

The taste enhancing effect of GMP and IMP was discovered in 1913 by Dr. Shintaro Kodama, a scientist at Tokyo University. Dr. Kodama isolated a close relative of IMP (histidine salt of inosinate) from dried bonito tuna, a traditional condiment in Japan. Since then, IMP has been found in a wide variety of fish and animal meats at

[33] "Save substantial sums of money" is the way Takeda U.S.A., Inc., lures customers for its product containing GMP and IMP in an advertisement in *Food Product Development,* May 1971, page 49.

levels up to 0.3 percent, and GMP has been found at high concentrations in some kinds of mushrooms and several species of fish. They contribute greatly to the natural flavor of these foods. The IMP in meat (muscle) forms from ATP, the "energy" molecule in living organisms, when the ATP breaks down after an animal dies. In properly aged meat most of the ATP has changed to IMP; if aging is too prolonged, the IMP decays to inosine and hypoxanthine and the flavor is lost.

GMP and IMP have been sold commercially as food additives in the United States for about ten years. You will find them in powdered soup mixes, ham and chicken salad spread, sauces and canned vegetables. Levels of usage vary from 0.003 to 0.05 percent. Manufacturers often use them together with monosodium glutamate, because of a synergistic action that exists between the three chemicals. GMP and IMP cannot be used in many moist foods, because enzymes in the food slowly convert the flavor enhancers to inert substances.

These two additives are not quite interchangeable with MSG. IMP and GMP are about ten to twenty times as potent and forty times as expensive, and have a subtly different taste effect. But like MSG, these flavor enhancers impart the impression that the food containing them has extra thickness and "meatiness." These qualities enable manufacturers to reduce the amount of expensive meat extract in their products. The use of disodium guanylate and inosinate will probably be limited until their cost declines.

Chemically, both GMP and IMP are related to the nucleic acids (DNA, RNA) of which our genes are composed, whereas MSG is an amino acid and a constituent of proteins. IMP and GMP are made from a natural product, yeast ribonucleic acid (RNA).

The way the body handles GMP and IMP is well understood, and both chemicals are safe. The IMP is degraded in the small intestine to hypoxanthine, while GMP is degraded to guanine. Both compounds are then absorbed by the body, converted to uric acid, and excreted. Persons who must avoid purines and uric acid, such as sufferers of gout and certain other genetic diseases, should not eat foods containing these additives.[34]

Ref.: C&EN p. 110; CRC-Ch. 13; *Food Tech. 18* 287, 298 (1964); Takeda, U.S.A., Inc., New York; Ajinomoto Company of New York; 21 CFR 121.1090, 121.1109.

EDTA[35]

Modern food manufacturing technology, which involves metal rollers, scrapers, blenders, and containers, guarantees that trace amounts of metal contamination will be present in food. These contaminants are undesirable, not only because they may be dangerous, but because they may impair the taste, odor, or appearance of food. Minute amounts of metal cause beverages to become cloudy, fruits and vegetables to discolor, and fats to spoil.

EDTA is extremely efficient at trapping the inevitable metal impurities. The structure of this molecule is such that positively charged metal ions—aluminum, copper, iron, manganese, nickel, zinc—are attracted to and trapped within a cage of negative charges. Manufacturers frequently use EDTA in combination with other antioxidants, such as BHT and propyl gallate, to make up a highly effective two-pronged attack on oxidation.

EDTA functions well in water and in oil-water mixtures. It does not dissolve in pure fats and oils so is not used in them.

[34] In gout, painful deposits of uric acid accumulate in joints.
[35] The full name of this additive is *e*thylene*d*iamine *t*etra*a*cetic acid; the calcium disodium and disodium salts of the acid are used in food.

You will find EDTA in many manufactured salad dressings, margarine, mayonnaise, and sandwich spreads in which it prevents the vegetable oil ingredient from going rancid. EDTA inhibits the metal-catalyzed browning of processed fruits and vegetables (potatoes, peas, corn, mushrooms, etc.) and prevents oxygen from destroying vitamin C in fruit juices. Canned shellfish—crabs, clams, and lobsters—usually contain large amounts of metal, often as high as 500 ppm, which promote discoloration and unpleasant odors and tastes; again, it's EDTA to the rescue. In beer, EDTA traps trace amounts of iron, which might otherwise cause a geyser of foam when the container was first opened, and copper, which causes clouding. Finally, soft drink companies use this versatile chemical to prevent artificial colors from fading.

Manufacturers are expected to add to a food only that amount of EDTA which is needed to trap metal impurities. Large excesses would combine with calcium, iron, and other nutrients and prevent them from being utilized by the body. EDTA is most often used at a concentration of 0.01 percent, but some products require as little as 0.0025 percent, others as much as 0.05 percent, depending on the amount of trace metal present.

Doctors used EDTA for medical purposes long before food chemists recognized its usefulness as a food additive. Physicians treat acute metal poisoning by giving patients large intravenous injections of EDTA. The EDTA captures the noxious metal ions in the blood, and the EDTA-metal complex is then excreted in the urine. This treatment is not 100 percent effective, but hundreds of lives are saved by it every year.

A second clinical application is to prevent the clotting of blood that is drawn for transfusions. The EDTA works by binding calcium, which is an essential component of the clotting reactions.

Occasional patients have suffered kidney damage from the injection of large doses of EDTA, but a detailed lifetime feeding study on rats, conducted by Dr. Bernard Oser, head of a private testing laboratory, indicated that 0.5 percent EDTA did not affect bones or teeth (which need calcium), growth, organs (including the kidneys), tumor incidence, or reproduction. A year-long study in dogs gave similar results. Metabolic experiments have shown that the body absorbs only about 5 percent of an oral dose and that this small amount is subsequently excreted in the urine.

Large doses of EDTA damage chromosomes in plant and cultured animal cells, but this damage appears to be due to a general disruption of cellular metabolism rather than to any intrinsic mutagenicity of EDTA.

Ref.: FAO(40A)-39; CUFP-15; CRC-Ch. 7; *J. Am. Med. Asso.* 174 263 (1960); *Tox. Appl. Pharm.* 5 142 (1963); *Fd. Cos. Tox.* 2 741 (1964); 21 CFR 121.1017, 121.1056.

Ergosterol   GRAS

Ergosterol is a natural steroid in yeast and mold that is converted by ultraviolet radiation to vitamin D. Irradiated ergosterol is added to milk to help people get an adequate amount of vitamin D.

Ferrous Gluconate

Ferrous gluconate is the iron salt of gluconic acid, a chemical which is present naturally in the body. Ferrous gluconate is used by the pharmaceutical industry as an iron supplement in vitamin pills and since about 1960 by olive growers as an artificial coloring.

The black olive, surprisingly, is a highly processed fruit. Olives destined to be black olives are immature at the time of harvest, and are straw yellow to light green in color and extremely bitter. Treating them with 1–2 percent lye (sodium or potassium hydroxide) destroys the bitter chemical constituent, and exposing them to air turns the skin brown. Variations in color from one olive to another are remedied by ferrous gluconate, which darkens the fruit to a uniform, jet-black color.

Treating olives with ferrous gluconate is perfectly safe from a health standpoint, but levels greater than approximately 40 parts per million may detract from the flavor. Olives purchased by Safeway Stores, Inc. are treated with 25–30 ppm gluconate, but it could not be determined if this level of treatment is standard in the industry. The Consolidated Olive Growers, California Canners and Growers, and the National Canners Association all declined to answer the author's inquiries. According to the Department of Agriculture, several processors have stopped using this additive, because they found that it sometimes makes olives unacceptably soft.

Cans of gluconate-treated black olives are deceptively labeled, because the function of the additive is not stated on the label. FDA regulations require all foods containing artificial coloring to reveal that fact on their labels. A complaint sent to the FDA by the author was rejected.

Ref.: *Encyclopedia Brittanica* ("olive"); pers. comm. from Safeway Stores, Inc., C. C. Graber Co., Drs. H. T. Hartmann and R. T. Vaughan (U. of California, Davis), FDA; 21 CFR 8.320.

Fumaric Acid   GRAS

Fumaric acid is a solid at room temperature, inexpensive, highly acidic (more acidic than citric acid, for instance), and does not absorb moisture readily. This set of qualities makes fumaric acid the ideal source of tartness and acidity in such dry products as gelatin desserts, pudding, pie fillings, candy, instant soft drinks, and leavening agents.

One drawback of fumaric acid is that it dissolves very slowly in cold water. For a time this property prevented food manufacturers from using it in the increasingly popular, powdered, cold beverage mixes. Recently, however, Monsanto chemists discovered that the acid would dissolve readily in cold water if it were mixed with 0.3 percent dioctyl sodium sulfosuccinate, a "wetting agent." If you survey the lists of ingredients on packages of powdered cold drinks, you will notice that "cold water soluble" fumaric acid is widely used.

Fumaric acid is an important metabolite and is present in every cell of the body. Biologists first examined the toxicity of the acid during World War II, when food manufacturers began to use fumaric acid in place of tartaric acid, which had become difficult to obtain from its manufacturers in the wine-producing countries of Europe. Experiments on a variety of mammals, including man, indicate that fumaric acid is harmless.

Ref.: C&EN-116; FAO(40A)-135; CRC-256, 639; CUFP-43.

Furcelleran   GRAS

Furcelleran is a vegetable gum whose composition and properties are similar to carrageenan's. Like carrageenan, this food additive is obtained from seaweed and is used by food manufacturers as a gelling agent in milk pudding and as a thickening agent in many foods.

Ref.: FAO(46A)-93.

Gelatin   GRAS

The first lesson in modern "cooking" that many children have is preparing gelatin dessert. Kids are often amazed when they add a brightly colored powder to water and see the solution turn into a firm gel. The magic ingredient is, of course, the protein called gelatin.

Gelatin is obtained from collagen, the most abundant protein in animals and the major constituent of tendons and ligaments (gristle in cooked meat). The structure of collagen may be likened to a many-stranded rope. When it is heated the collagen (rope) breaks up into gelatin molecules (the strands). When the gelatin is dissolved in hot water, which is then allowed to cool, the molecules react weakly with one another to form a gel. If you could see a gel under a microscope, it would look like a vast, tangled three-dimensional network of long, skinny molecules.

In addition to gelatin, powdered dessert mixes also contain sugar or dextrose for sweetness, fumaric or adipic acid for tartness, sodium citrate to control the acidity, artificial coloring, artificial or natural flavoring (depending on the brand), and in some brands and flavors a pinch of the preservative BHA to protect the flavoring.

Gelatin is not the only natural substance that can form a gel—agar, carrageenan, and other carbohydrates extracted from seaweed also can—but it is perfect for foods because its gel melts at body temperature. On the other hand, in hot climates gelatin desserts will melt unless they are refrigerated.

Food manufacturers sometimes use gelatin at low concentrations to thicken yogurt, ice cream, cheese spreads, and beverages. Gelatin is pure protein but has little nutritive value because it contains little or none of the essential amino acids.

Ref.: CRC-341.

Gluconic Acid   GRAS

Gluconic acid is occasionally used as a component of leavening in cake mixes or as an acid in powdered gelatin dessert and soft drink mixes. This chemical occurs naturally in the human body and is quite safe. See "ferrous gluconate."

Ref.: FAO(40A)-139.

Glycerin (glycerol)   GRAS

Glycerin—also called glycerol—is a clear, thick liquid that manufacturers add to foods to maintain a certain moisture content and prevent foods from drying out and becoming hard. You will find it in marshmallows, candy, fudge, and baked goods in amounts ranging from 0.5 to 10 percent. Manufacturers also use glycerol as a solvent for oily chemicals, especially flavorings, that are not very soluble in water.

Glycerin is safe. It forms the backbone of fat and oil molecules and is a familiar chemical to the body. The body uses glycerin either as a source of energy or as a starting material in making more complex molecules. Combining glycerin with three molecules of nitric acid results in nitroglycerin; fortunately, this reaction does not occur in living organisms.

Ref.: CRC-452; CUFP-265.

Glycine (aminoacetic acid)

Glycine is an amino acid that is present in all proteins. It is especially abundant in collagen (20–35 percent), the protein that makes up much of the body's connective tissue. Bottlers add glycine (0.2 percent) to artificially sweetened soft drinks to mask the bitter aftertaste of saccharin.

Ref.: Wall Street Journal, May 19, 1970; 21 CFR 121.4002.

Guar Gum   GRAS

Guar gum is one of the most widely used, cheapest, and most effective of the vegetable gum stabilizers. A given weight of guar will form a thicker solution than any other stabilizer. Guar is also unusual among gums because it dissolves in cold water. As an added bonus, a guar gum solution will turn into a rubbery gel if borate is added as a cross-linking agent.

The guar plant, which resembles the soybean plant, is grown mainly for use as cattle feed in the United States (Texas and Oklahoma), Pakistan, and India. The endosperm of the seed of this plant is virtually pure guar. The remainder of the seed is a valuable source of protein for livestock. In fact, so valuable is the protein of guar seeds that India forbids and Pakistan discourages the export of whole seeds.

Guar gum serves as a thickening agent in beverages, ice cream, frozen pudding, and salad dressing. Manufacturers also use it to increase the resiliency of doughs and batters and in the production of artificial whipped cream.

Biochemists have shown that guar gum is digestible; a long-term animal feeding study is currently in progress.

Ref.: FAO(46A)-100; *Tox. Appl. Pharm.* 5 478 (1963).

Gum Arabic   (acacia gum, gum senegal)   GRAS

Gum arabic is obtained from a short, middle-eastern tree, the acacia, in which it serves as a wound sealant. The gum oozes out and is collected from trees whose bark has split open because of drought, microbial infection, or intentional cuts. American firms obtain virtually all their gum arabic from the Sudan.

Archeologists have discovered that the Egyptians used gum arabic four thousand years ago in paint colors. These days, food processors use it to prevent sugar crystals from forming in candy, to help citrus oils dissolve in drinks, to encapsulate flavor oils in powdered drink mixes, to stabilize the foam in beer, and to improve the texture of commercial ice cream. In 1970 U.S. companies imported twenty-nine million pounds of gum arabic, making it one of the two most widely used vegetable gum thickeners.

Gum arabic is unusual among vegetable gums because it is very soluble in water; solutions become viscous only when they contain 10–20 percent gum. Another unusual property of gum arabic is that it is completely digested by rodents (rat, guinea pig) and, presumably, man. The ability of gum arabic (or impurities associated with it) to cause cancer, birth defects, or mutations has never been studied. We cannot consider this additive safe until such tests are done.

Ref.: FAO(46A)-97.

Gum Ghatti   GRAS

Gum ghatti is a vegetable gum, which, like karaya and gum arabic, is produced as a wound sealant by a tree indigenous to India. Food manufacturers use it to keep the oil and water ingredients from separating out into two layers in such products as salad dressing and butter-in-syrup. Approximately 800,000 pounds were imported in 1963. No toxicological studies have been conducted, so the safety of gum ghatti cannot be judged.

Gum Guaiac   GRAS

Gum guaiac is a greenish brown resin obtained from the tropical guaiacum tree. Foodmakers occasionally use it to prevent oil-containing foods from spoiling. This antioxidant has not been adequately tested and should not be permitted in foods.

Ref.: FAO(31)-65.

Hesperidin   GRAS

Hesperidin is a bioflavonoid that occurs in the pulp of citrus fruits. Aside from a minor use as a flavoring, hesperidin is used therapeutically to strengthen blood capillaries. It is sometimes called vitamin P.

Ref.: Ann. N.Y. Acad. Sci. 61 (3) (1955); pers. comm. from Sunkist Growers, Ontario, Calif.

Hydrolyzed Vegetable Protein (HVP) GRAS

Foodmakers use hydrolyzed vegetable protein (HVP) to bring out the natural flavor of food. HVP consists of vegetable (usually soybean) protein that has been chemically degraded to the amino acids of which it is composed. Some of the products in which you will find it are instant soups, beef stew, frankfurters, gravy and sauce mixes, and canned chili.

Hydroxylated Lecithin GRAS

Hydroxylated lecithin is manufactured by treating soybean lecithin with peroxide. The food industry uses it as an emulsifier and antioxidant in baked goods, ice cream, and margarine.

Scientists at Central Soya Company have fed this additive to small groups of laboratory animals for brief periods of time and did not detect any adverse effects, but thorough studies are needed to prove that hydroxylated lecithin does not cause cancer, mutations, birth defects, or other problems. The FAO/WHO Expert Committee on Food Additives recently reported that "hydroxylated lecithin has not been adequately studied from the toxicological point of view and there are not satisfactory data available on which an evaluation could be based." Until "satisfactory data" are available, the American public should not be eating this synthetic chemical.

Ref.: FAO(46A)-127; pers. comm. from FDA, Central Soya, Ft. Wayne, Ind.

Imitation Beef and Chicken Flavors

Imitation beef and chicken flavors contain:

> hydrolyzed vegetable protein
> monosodium glutamate
> sugars
> vegetable fat
> amino acids
> disodium inosinate
> disodium guanylate (chicken flavor only)
> modified starch

These mixtures can be used alone or, for fuller flavor, in combination with spices and real beef or chicken extracts. Their use allows manufacturers to skimp on the more expensive and nutritious natural ingredients.

Ref.: Chas. Pfizer & Co., Data Sheet No. 638.

Invert Sugar

Invert sugar is a 50-50 mixture of two sugars, glucose and fructose, that is used by candy and other manufacturers because it is sweeter, more soluble, and crystallizes less readily than sucrose (ordinary table sugar). Invert sugar forms when sucrose is split in two by an enzyme (invertase) or an acid. It is called "invert" because a solution containing the 50-50 mixture and a solution containing sucrose rotate the plane of polarized light in opposite directions (consult a physics textbook or professor for an explanation).

Karaya Gum   GRAS

Karaya gum is a complex carbohydrate that manufacturers use as a thickening or stabilizing agent in foods. The only source of karaya is the sterculia tree which grows in India. To obtain the gum, trees are tapped, causing the extremely viscous gum to exude slowly out of the wound. Tear-shaped masses of gum, which weigh up to five pounds, are harvested from the tree trunks. In 1970 American companies imported approximately eight million pounds.

Manufacturers use karaya gum to prevent oil from separating out of whipped products and salad dressing and to prevent fat from separating from meat and juices in sausages. It improves the texture of manufactured ice cream and sherbet by preventing large ice crystals from forming. Karaya is proving to be a cheap substitute for tragacanth gum.

What distinguishes karaya from other natural gums is its exceptional proclivity for water, in which it can expand to one hundred times its original volume. This property makes it an effective bulk-type laxative. The stickiness of karaya solutions enables it to be used in hair wave-set products. Considerably more karaya is used in pharmaceuticals and cosmetics than in foods.

Good long-term feeding studies need to be done, with particular attention paid to the effect of karaya on the absorption of nutrients. The ability to function as a laxative indicates that very little of this gum is digested and absorbed, but the fact that some persons are allergic to karaya indicates that some is absorbed. The metabolism of karaya should be thoroughly investigated.

Ref.: FAO(46A)-102; *J. Nutr.* 36 27 (1948); *J. Am. Med. Assoc.* 114 747 (1950).

Lactic Acid   (calcium lactate)   GRAS

The significance of lactic acid for animals and plants has been nicely described by Dr. Myron Brin, a professor of nutrition at the University of California at Davis:

> To a biologist, lactic acid is consonant with life, since the presence of this acid is a hallmark for energy metabolism

for virtually every living organism . . . It is ubiquitous in distribution, being found wherever life persists.

Lactic acid is also found where many food industries persist. Food manufacturers use the acidic properties of lactic acid to inhibit the spoilage of Spanish-type olives, to adjust the acidity in cheese-making, and to add a note of tartness to frozen desserts, carbonated fruit beverages, and other foods. Calcium lactate, a non-acidic salt of lactic acid, inhibits the discoloration of fruits and vegetables and improves the properties of dry milk powders and condensed milk.

As a consequence of its chemical structure, lactic acid exists in two forms: L-lactic acid, which is the form normally encountered in nature, and D-lactic acid, which is the mirror image of the first form (like right-handed and left-handed gloves). Commercial lactic acid is a 50-50 mixture of the two forms. Animals have not been extensively exposed to D-lactic acid, but they are able to metabolize it, though not so readily as L-lactic acid. Unmetabolized lactic acid is excreted in the urine. The FAO/WHO Expert Committee on Food Additives recommends that because the metabolism of D-lactic acid has not been studied in infants, they should not be exposed to it unnecessarily.

Ref.: FAO(40A)-144; M. Brin in *Ann. N.Y. Acad. Sci. 119* 108 (1965) (whole issue on lactic acid).

Lactose GRAS

Lactose, a carbohydrate that is found only in milk, is Nature's way of delivering calories to infant mammals. Human milk contains about 7 percent lactose and cow's milk contains 5 percent. Milk turns sour when bacteria convert lactose to lactic acid.

The intestine secretes an enzyme that splits lactose into glucose and galactose, two sugars that the body absorbs and metabolizes. The only persons who may be harmed by lactose are the rare—perhaps one out of twenty thousand—individuals who suffer from galactosemia, a genetic disease. Galactosemics lack the enzyme that metabolizes galactose. In the absence of the enzyme, the galactose that forms from lactose accumulates in the blood and causes enlargement of the liver, cataracts, and mental retardation. Proper treatment for this disease entails simply the avoidance of milk. Interestingly, galactosemia is mainly a childhood disease. The reason for this is that children have only one enzyme that metabolizes galactose, and this enzyme is missing in galactosemics; adults develop a second enzyme which puts the galactose to good use.

Food manufacturers use lactose in whipped topping mix, fortified breakfast pastry (toaster tarts), and other foods as a slightly sweet source of carbohydrate. Lactose is one sixth as sweet as table sugar.

Lecithin   GRAS

Lecithin occurs in nearly all animal and plant tissues and is an important constituent of the human body. The average person consumes one to five grams (up to one sixth of an ounce) of lecithin in his daily diet. Food manufacturers use lecithin as an antioxidant and emulsifier.

Pure lecithin is an important source of only one nutrient, choline, but the lecithin that is sold commercially contains many valuable "impurities," including vitamins B-1, B-2, and E, inositol, niacin, and biotin.

The lecithin used as a food additive is obtained almost exclusively from the soybean, as it has ever since that versatile plant was introduced into the United States in the 1920s. Prior to that time it was obtained from egg yolk. In an era of sky-high prices, soy lecithin costs a remarkably low twelve cents a pound. The price is so low because lecithin is a waste product of the soybean industry. Before shippers will transport soy oil, they require that the lecithin be removed because it absorbs moisture and forms difficult-to-remove deposits in ocean-going tankers. Far more lecithin is removed from the oil than is used commercially so there is a glut of lecithin on the market, thereby driving down the price. Lecithin obtained from egg costs approximately $20 per pound.

In living animals and plants lecithin protects polyunsaturated fats from attack by oxygen and helps maintain biological membranes. The food industry takes advantage of the antioxidant action and adds up to 0.5 percent lecithin to margarine, shortenings, and oils to retard spoilage and rancidity. Lecithin is a relatively weak antioxidant compared to the synthetic compounds BHA, BHT, and propyl gallate, but it is certainly safer.

Foodmakers use lecithin primarily as an emulsifier in margarine, chocolate, ice cream, and baked goods to promote the mixing of oil (or fat) and water. It is one of the few natural, edible substances that has a strong emulsifying effect. Like other emulsifiers, lecithin works by lowering the surface tension of water.

In margarine lecithin serves several interesting purposes. Already mentioned was its antioxidant effect; in addition to retarding spoilage, it protects vitamin A (beta carotene). Because lecithin attracts moisture, it prevents the leakage of water from margarine ("weeping"). Its effect on surface tension reduces spattering in a frying pan. When the margarine is used in bread and rolls, the lecithin exerts a tenderizing effect.

In baked goods lecithin helps the shortening mix with the other dough ingredients, retards the crystallization of starch, and stabilizes air bubbles in cake batter. The end result is bread and rolls that are more tender and stay fresh longer and fluffier cakes.

Adding 0.3 percent lecithin to chocolate enables manufacturers to reduce the cocoa butter content from 36 to 32 percent. Most manufacturers employ this trick because cocoa butter is the most expensive ingredient in chocolate. The Hershey Foods Corporation claims that this substitution has an insignificant effect on the taste of chocolate.

The lecithin that is present in egg yolk serves as a natural emulsifier in many foods. In mayonnaise, for instance, it stabilizes the vinegar-vegetable oil mixture, preventing these ingredients from separating into two layers. The fact that egg lecithin is associated with protein makes it an even more powerful emulsifier than pure lecithin.

In summary, lecithin is nutritious and nontoxic and serves several important technological functions as a food additive.

Ref.: FAO(35)-151; *ECT 12* 343; West, E. S., and Todd, W. R., *Textbook of Biochemistry,* Macmillan, N.Y., p. 147 (1961); pers. comm. from Hershey Foods Corp.

Locust Bean Gum (Carob seed gum; St. John's bread)
GRAS

Locust bean gum, or St. John's bread as it is sometimes called, is obtained from the endosperm of the bean of the carob tree, a Mediterranean species. In ancient Egypt the gum was used as an adhesive in mummy bindings. Today it serves as a stabilizer in foods and, when taken in greater quantity, as a gentle laxative. Spain, Portugal, Italy, and Greece fill America's demand for this gum (eleven million pounds in 1970).

Locust bean gum is used in food to improve the texture and freeze-melt characteristics of ice cream; thicken salad dressing, pie filling, and barbecue sauce; make softer, more resilient cakes and biscuits when used as a dough additive; and increase the palatability of carra-geenan gels by decreasing their brittleness and melting temperature.

That locust bean gum has a laxative action indicates that it is not absorbed to any great extent by the body. However, impurities that may be present in this natural product could be harmful. Animal feeding studies must be conducted before this substance can be considered safe.

Ref.: FAO(46A)-99; *Am. J. Dig. Dis. 18* 24 (1951).

Malic Acid GRAS

Malic acid sometimes goes by the name "apple acid," because large amounts occur in apples (0.4 to 0.7 percent in apple juice). Other fruits contain smaller amounts of this tart acid. The change in flavor that occurs as a fruit ripens is due partly to a decrease in the malic acid content and to an increase in the sugar content. Malic acid is an important metabolite and is present in all living cells.

The food industry has used malic acid for over forty years as an acidulant and flavoring agent in fruit-flavored drinks, candy, lemon-flavored ice-tea mix, ice cream, and preserves.

Because of its chemical structure, two mirror-image forms of malic acid exist: the D form and the L form. L-malic acid, the form that occurs in natural products, is metabolized routinely. Commercial malic acid, however, is a mixture of the D and L forms. While adults can probably utilize the D-malic acid, it is not known whether infants can utilize it. Therefore, it is important that synthetic malic acid not be added to baby food.

Ref.: FAO(40A)-149; C&EN-117; CUFP-44, 144; *Fd. Cos. Tox. 7* 103 (1969).

Maltol, Ethyl Maltol

Food manufacturers use maltol and ethyl maltol to enhance the flavor and aroma of fruit-, vanilla-, and chocolate-flavored foods and beverages the way they use MSG to bring out the flavor of meat. Small amounts of maltol occur naturally in bread crust, coffee, and chicory and as a degradation product in heated milk, cellulose, and starch. It has been available commercially since 1942.

Chemical manufacturers frequently synthesize families of compounds that are closely related to substances that they manufacture. The purpose of creating these new substances is to learn whether a close relative of a widely used compound might have greater potency, stability, safety, or lower cost. Using this technique, one manufacturer synthesized ethyl maltol and discovered that it was four to six times as strong a flavor enhancer as its close relative, maltol.

Maltol and ethyl maltol are used in gelatin desserts, soft drinks, ice cream, and other foods that are high in carbohydrate. Levels used range from 15 to 250 parts per million for maltol and 1 to 50 ppm for ethyl maltol. Ethyl maltol can be used in diet foods to mask saccharin's bitter aftertaste and in ordinary foods to permit a lower sugar content.

The manufacturers of maltol have not done experiments to verify their assumption that the additive does not cause cancer, birth defects, or mutations.

Ethyl maltol has been tested more extensively but still not adequately. In a lifetime feeding study on rats, Pfizer Chemical Company biologists found that food containing 0.4 percent ethyl maltol did not cause tumors or impair growth. Additional tests of shorter duration on rats and dogs also indicated safety. Ethyl maltol is completely absorbed by the body; following absorption, it is modified in the liver (to a sulfonate or glucuronide) and then excreted in the urine.

Both maltol and ethyl maltol should be the subject of careful carcinogenicity and teratology studies in several species of animals.

Ref.: FAO(44A)-56; *Tox. Appl. Pharm. 15* 604 (1969); Chas. Pfizer & Co. Data Sheet No. 635; 21 CFR 121.1164.

Mannitol   GRAS

Sweetness, disdain for moisture, poor absorption by the body, and diuretic action are some of the properties that make mannitol an interesting and useful chemical. This versatile substance occurs naturally in the manna ash tree, in which it helps to close wounds, and in seaweed and microorganisms. Commercial quantities of mannitol are made by synthesis from sugar.

Perhaps the most familiar use of mannitol is as the "dust" on chewing gum. The dust's main function is to prevent gum from absorbing moisture and becoming sticky. Mannitol's sweetness—it is two thirds as sweet as sugar—is an added bonus.

Although mannitol is chemically quite similar to sugar, few organisms are well adapted to use it as a source of energy. The human digestive tract, for instance, absorbs only about two thirds of a given dose of mannitol, and even some of that is excreted unchanged in the urine. Approximately 50 percent is used by the body, which means that mannitol generates only half as many calories as an equal weight of sugar. Thus, mannitol may be used as a sweetener in low calorie foods.

Mannitol is frequently used as the sweetening agent in noncariogenic ("sugarless") chewing gum, because bacteria in the mouth have an even harder time digesting it than do humans. Tooth decay is caused by acid produced by oral bacteria as they metabolize sugar.

Physicians sometimes call upon mannitol to help clean out the body's plumbing. Moderate amounts (about one ounce) have a laxative effect because it is poorly absorbed. When mannitol is injected or fed intravenously, it is totally eliminated by the kidneys and has a powerful diuretic effect. The increased production of urine helps eliminate poisons circulating in the bloodstream. The diuretic effect also helps prevent kidney shutdown in patients undergoing certain major operations; this technique has saved many lives.

Mannitol has been used for decades to "cut" heroin. According to a Justice Department spokesman, a typical $5 bag of heroin purchased in New York City contains eight times as much mannitol as heroin. (Most heroin also contains a dash of quinine, which mimics some of heroin's physiological effects and also adds to the bitterness of the dope. This twofold effect leads addicts to believe that they are using high quality heroin.)

People have used mannitol for centuries as a sweetening agent without adverse effect. The body converts mannitol to a sugar, which in turn is converted to energy. These considerations, plus a variety of short-term tests, indicate that mannitol is safe.

Ref.: FAO(40A)-160; CRC-Ch. 11; Atlas Chemical Company's "Atlas Products for the Food and Beverage Industry"; *Chemical & Engineering News,* January 17, 1972, page 13.

Meat Tenderizer   GRAS

The average life span of medieval magicians would probably have been increased greatly had they spent less time trying to turn lead into gold and more time trying to make tough meat more tender.

The best procedure that butchers, if not magicians, developed to improve the texture of meat was to age it.

Scientists have discovered that during aging, natural protein-digesting enzymes (proteases) in the meat gently loosen up muscle fibers that had contracted in *rigor mortis*.

Food technologists took a cue from nature and found that meat could be artificially tenderized simply by treating it with proteases, which are readily purified from plants, animals, or microorganisms. This trick was discovered ages ago by natives of Mexico, South America, and the South Sea Islands who traditionally tenderized meat by wrapping it in papaya leaves, a rich source of enzymes.

Family cooks tenderize steak by sprinkling a small amount of tenderizer on it shortly before cooking. A more efficient method that some meatpackers use is to inject proteases into a steer's bloodstream shortly before the animal is slaughtered. The blood carries the enzymes to all parts of the animal, resulting in a better job than the superficial tenderizing at home. Federal regulations require that meat treated in this way be appropriately labeled. In practice, however, only the carcass is stamped "tenderized"; individual steaks in butcher shops and restaurants will probably not be labeled.

Meat tenderizing proteases are extracted from plants or microorganisms. The ones derived from plants—bromelain (pineapple[36]), papain (papaya), ficin (fig)—act most effectively on gristle (connective tissue composed of collagen and elastin), although they also tenderize the muscle portion of the meat. Bacterial and fungal proteases (from *Aspergillus* and *B. subtilis*) have the opposite priorities.

While I could not recommend eating spoonfuls of meat tenderizer straight from the jar, it is normally harm-

[36] Fresh pineapple cannot be used in gelatin desserts because the bromelain dissolves the gel, which is made of protein (gelatin).

less. The enzymes are destroyed when the meat is cooked and even if they were not destroyed, our mouth and throat are amply protected by saliva and mucus.

Meat tenderizer preparations consist mostly of seasoning; a small amount of edible fat, such as calcium stearate, is added to prevent caking and to keep down the dust.

Ref.: CRC-62, 87; *Enzymes in Food Processing*, G. Reed, Academic Press, N.Y., 1966; pers. comm. from Swift & Co., Chicago, and Adolph's Food Products Mfg. Co., Burbank, Calif.

Mono- and Diglycerides and related substances

The words "mono- and diglycerides" are familiar to every red-blooded, label-reading American. This chemical additive makes bread softer and prevents staling (by preventing the starch from crystallizing), improves the stability and taste of margarine, makes cakes fluffier (by helping generate and trap air bubbles), decreases the stickiness of caramels, prevents the oil in peanut butter from separating out . . . the list goes on and on. The average American consumes over a half pound of mono- and diglycerides as food additives annually.

Mono- and diglycerides are safe and wholesome. In fact, they comprise about 1 percent of normal food fat and are part of our normal diet. Moreover, the bulk of normal food fats and oils is composed of triglycerides, which our body's digestive system converts to mono- and diglycerides. Cooking, also, converts a large fraction (up to 40 percent in bread) of fats and oils to mono- and diglycerides. The mono- and diglycerides derived from either food fat or food additives are absorbed into intestinal cells where they are largely converted to triglycerides. The triglycerides then pass into the bloodstream.

While mono- and diglycerides are completely harmless

from a toxicological point of view, they may displace more nutritious ingredients from some foods. Thus, "modern" peanut butter contains up to 10 percent mono- and diglycerides to keep the oil from separating, which means that it is approximately 10 percent less nutritious than "old-fashioned" peanut butter made solely from peanuts.

Food technologists have synthesized useful new substances by reacting mono- and diglycerides with other compounds. Bakers use one of these, ethoxylated mono- and diglycerides[37] as a dough conditioner to improve the baking characteristics and texture of yeast-raised bakery goods. Judging from reports on file at the FDA of rat and dog studies, none of which studies was longer than ninety days' duration, this substance is partially digested and absorbed into the blood. Ethoxylated mono- and diglycerides should be studied in a complete battery of tests on mice, rats, and dogs to see if they cause cancer or birth defects. This food additive should be banned until scientific tests demonstrate a lack of hazard.

A second derivative is sodium sulfoacetate mono- and diglycerides. This chemical was developed as a synthetic substitute for lecithin and, like the more nutritious lecithin, is used as an antispattering agent and emulsifier in margarine. As far as could be determined this chemical has not been adequately tested. If this additive is better technologically than lecithin, the difference is probably marginal. Most margarine manufacturers find lecithin perfectly adequate. Because sodium sulfoacetate mono- and diglycerides has not been well tested and because it offers no significant benefit to the consumer, it should not be allowed as a food additive.

Acetylated mono- and diglycerides is one of the more unusual derivatives of the parent compound. Unlike its

[37] Also called polyoxyethylene (20) mono- and diglycerides.

greasy relatives, this substance is a highly flexible and wax-like solid at room temperature. You are likely to encounter it in the coatings on jelly beans and chocolate-covered ice cream bars. It is completely digestible and harmless.

Three other harmless derivatives are the lactylated, citrated, and succinylated mono- and diglycerides. All of these normally serve as emulsifiers, although the citrate derivative also retards spoilage of oils (see "citric acid").

The final derivative, diacetyl tartaric acid ester of mono- and diglycerides, is used by the baking industry. When this emulsifier was introduced, about 1950, only small-scale feeding experiments had been conducted. These studies are crude and incomplete by today's standards and need to be updated so that we can be certain that this synthetic compound does not cause cancer, birth defects, or other harm. The FDA should suspend the use of this additive until its manufacturers can prove that it is safe.

Nice as it would be if we could avoid the untested derivatives of mono- and diglycerides, we can't. The reason for this is that they are most commonly used in bread and rolls and, as explained in Chapter 3, the ingredients of these products are never listed on the wrapper.

Ref.: Mono- and diglycerides: C&EN-120; CUFP-22; NAS-NRC Pub. No. 1271; CRC-Ch. 10; FAO(35)-145-151; GRAS.

Ethoxylated mono- and diglycerides: pers. comm. from FDA; 21 CFR 121.1221.

Sodium sulfoacetate mono- and diglycerides: *ECT 12* 359; FAO/WHO Expert Committee on Food Additives, tentative specifications.

Acetylated mono- and diglycerides: FAO(40A)-71; *J. Am. Oil Chem. Soc. 29* 11 (1952); *J. Nutr. 58* 113, *59* 277 (1956); 21 CFR 121.1018.

Diacetyl tartaric acid ester of mono- and diglycerides: FAO(40A)-90; *J. Am. Pharm. Asso., sci. ed. 39* 275 (1950); GRAS.

Other derivatives: FAO(40A).

Monosodium Glutamate (MSG)   GRAS

Monosodium glutamate is the sodium salt of glutamic acid, an amino acid, and is one of the building blocks of which protein molecules are made. It is present in all proteins and many foods.

MSG was just another amino acid until 1908. In that year Dr. Kikunae Ikeda, a chemist at Tokyo University, discovered that MSG had an extraordinary ability to intensify the flavor of protein-containing foods and was the active component of soy sauce and sea tangle, traditional oriental condiments. Japanese chemical firms soon began producing MSG commercially, selling it to both food manufacturers and consumers. Persons all around the world now use MSG as freely as table salt to enhance the flavor of meat, soup, seafood, poultry, cheese, and sauces. Manufacturers use MSG to accentuate the meatiness of a food (and possibly to reduce the amount of expensive meat or meat extract in a food).

Americans consumed about 45 million pounds of MSG in 1967. Chemical companies produce 90 percent of the MSG by the bacterial fermentation of sugar; they obtain the remainder from plant proteins rich in glutamic acid. Japanese firms produced 140 million pounds of MSG in 1965.

The first indication that MSG did anything but improve the flavor of food came in 1968 when a doctor of Chinese extraction, Ho Man Kwok, discovered what he dubbed "Chinese Restaurant Syndrome (CRS)." Approximately twenty to thirty minutes after beginning a meal at a Chinese restaurant, Dr. Kwok experienced a burning sensation in the back of his neck and in his forearms, tightness of the chest and headaches. With exemplary scientific vigor several New York doctors followed up Kwok's observations by eating three meals a day, sampling a wide variety of dishes, at their favorite Chinese restaurant. Many meals later they traced the cause of CRS to the soups. Back in the kitchen they discovered that the cook routinely added great quantities of MSG to the soup. Subsequent laboratory experiments confirmed the existence and cause of CRS and showed that some persons were ten or more times as sensitive as others.[38] Men appeared to be more sensitive than women. Chinese Restaurant Syndrome is readily produced by intravenous injection of small amounts of MSG. Doctors believe that CRS is only a minor nuisance and does not cause any permanent damage.

Many foods besides soups in Chinese restaurants contain MSG, but due to a curious combination of circumstances few others cause CRS. Chinese soups that contain large amounts of MSG are so potent because persons frequently go to Chinese restaurants on empty stomachs, and because soup is the first course of a meal. These factors ensure that a large amount of MSG is rapidly and completely absorbed into the blood, which carries it to nerve endings where it probably exerts its effects.

A more serious difficulty with MSG was brought to the public's attention in 1969 by Dr. John Olney of the Washington University School of Medicine in St. Louis. He found that feeding large amounts[39] of MSG to infant mice destroyed nerve cells in the hypothalamus, the region of the brain that controls appetite, body temperature, and other important functions. The identical effect was seen when MSG was injected under the skin of mice and of a monkey. When massive brain damage occurred, the baby mice grew up to be short and fat, and their coats,

---

[38] Not every experiment has demonstrated the existence of Chinese Restaurant Syndrome.

[39] Doses of 0.5 g/kg body weight and above.

livers, uteri, and ovaries were visibly affected. The nature of the effects on the infant monkey, Dr. Olney wrote,

". . . when a small percentage of its brain cells were being destroyed is evidence of a subtle process of brain damage in the developmental period which could easily go unrecognized were it to occur in the human infant under routine circumstances."

A second effect of feeding large amounts of MSG to newborn rats and mice was injury to the retina of the eye. Fortunately, though, the human eye is further developed at the time of birth than is the rodent eye and, therefore, is less susceptible to injury.

Baby food manufacturers in the United States had been adding MSG to their meat- and vegetable-containing products for many years. The manufacturers admitted that the function of MSG was to make baby food more palatable to mothers, and not to make the product more nutritious or tastier for the infant. But they maintained that there was far too little of the compound in a jar to harm a baby. Olney's experiments, however, suggested that a toxic amount of MSG might be contained in as few as four jars of food—a margin of safety too small for most parents' comfort. Moreover, viral infections, retarded development of enzyme systems, an empty stomach, and metabolic variations might cause many babies to be supersensitive. It took massive public pressure and the glare of publicity to force the baby food companies to stop adding MSG to their products as of October 1969. The industry did not recall food made prior to that date, however, and one and a half years later MSG-containing baby food could still be found on grocery store shelves. That industry finally capitulated should not obscure the way it abrogated its responsibility by not testing an ingredient on infant animals before using it in a product and by not removing it at the first sign of danger. A simlar situation appears to hold true for carrageenan, which s used in infant formula.

The effect of MSG on developing fetuses (in contrast o baby animals) began to receive study only after the baby food scare. No brain damage or other harm has been observed in recent experiments on rabbits and chickens (doses as high as 2.5 g/kg per day), but the retinas were not examined. The human placenta complicates matters by concentrating MSG and exposing the fetus to twice the concentration as the mother. Taking the placenta's effect into account, a pregnant woman is still unlikely to consume enough MSG to endanger her unborn child. But until the chemical has been more fully studied, pregnant women should take care not to consume gross amounts of MSG.

Lifetime feeding tests on rats and mice confirmed the expectation that MSG, an important metabolite and dietary element, does not cause tumors.

Persons who must restrict their intake of sodium use monopotassium glutamate. It has the same gustatory and biological effects as MSG.

Ref.: FDA Report on Monosodium Glutamate (November 17, 1969); NAS-NRC Food Protection Committee Report (July 1970), "Safety and Suitability of Monosodium Glutamate for Use in Baby Foods"; C&EN-109 (see extra story in reprint of article); *Nut. Rev. 28* 124, 158 (1970); *Science 170* 549 (1970); *Nature 27* 609 (1970); *Science 163* 826 (1969).

Oxystearin

Oxystearin is a modified fatty acid that manufacturers add in amounts up to 0.125 percent to vegetable oils to prevent them from clouding up in the refrigerator. Proctor & Gamble has sponsored lifetime feeding, metabolic and reproduction studies on animals that indicate that oxystearin is safe.

Ref.: FAO(46A)-155; 21 CFR 121.1016.

Parabens GRAS

Parabens is the nickname for the methyl, propyl, and heptyl esters of parahydroxybenzoic acid. These dreadful sounding chemicals are used as preservatives and are closely related to sodium benzoate. Parabens are more versatile than benzoate because they can prevent bacteria and mold from growing in almost all foods; benzoate is effective only in acidic foods. Parabens have also been used for many years in pharmaceutical products, where their relatively high cost is acceptable.

The most commonly used parabens, the methyl and propyl esters, have been fairly well tested in rats and dogs. The body absorbs parabens and converts them to parahydroxybenzoic acid, which is coupled to another substance (glucuronic acid, sulfate or glycine) and then excreted. Dr. C. Matthews and his colleagues at Emory University did not detect any adverse effects on the growth, life span or internal organs of rats that ate food containing 2 or 8 percent parabens for ninety-six weeks. An experiment conducted by Swedish biologists showed that parabens do not cause birth defects. Dogs were not harmed by eating for one year food that contained 3 percent parabens. Additional tests are needed on other species of animals before parabens can be considered safe.

Heptyl paraben acts as a preservative in beer (12 ppm). Studies conducted by scientists at Hazleton Laboratories show that this chemical does not cause cancer or interfere with the reproduction of rats.

Ref.: FAO(31)-30-37; FAO(40A)-22-27; CRC-142-151; *Acta. Paediat. Scand.* 54 43 (1965); *J. Am. Pharm. Asso., sci. ed.* 45 260, 268 (1956); pers. comm. from FDA; GRAS (up to 0.1 percent).

Pectin   GRAS

Pectin is a carbohydrate that strengthens cell walls in citrus fruits, apples, beets, carrots, and other fruits and vegetables. Fruits get soft as they ripen because the pectin breaks down to a soluble form.

Pectin forms gels that are the basis of fruit jellies. Jelly manufacturers add purified pectin to their products when the fruits contain too little natural pectin to form good gels. You may also find it used to thicken barbecue sauce, cranberry sauce, canned frosting, and yogurt.

Ref.: ECT 14 636.

Polysorbate 60, 65, 80[40]

Manufacturers have been adding polysorbate emulsifiers to processed foods since the 1940s. Like other emulsifiers, the main effect of polysorbates is to enable oil to remain dispersed in water. This action improves the texture and promotes "dryness" in frozen desserts, makes bread, rolls, and doughnuts more tender and keeps them from going stale, prevents oil from separating out of artificial whipped cream, helps nondairy coffee whiteners dissolve in coffee, keeps dill oil dissolved in bottles of dill pickles, and keeps flavor oils dissolved in candy, ice cream, and beverages. The polysorbates are employed at concentrations of 0.01 percent. In most of their functions, polysorbates are considerably more potent (on a weight basis) than fat or mono- and diglycerides. The differences between polysorbate 60, polysorbate 65, and polysorbate 80 are technical.

During the 1950s there was considerable controversy about the safety of the polysorbates. In some experi-

---

[40] The full names of these additives are polyoxyethylene-(20)-sorbitan monostearate, tristearate, and monooleate. They are sometimes called Tween 60, Tween 65, Tween 80 (trademarks of Atlas Chemical Industries, Inc.).

ments, rats and hamsters whose diets contained 25 percent polysorbate developed bladder stones, died prematurely, lost their tails, and suffered sundry other side effects. When hamsters ate food containing 5 percent polysorbates, they suffered diarrhea, early death, changes in the kidney, liver, and large intestine, and retardation of growth.

The FAO/WHO Expert Committee on Food Additives examined the experiments casting doubt on the polysorbates and gave the additives a clean bill of health. The Committee ascribed most of the adverse effects to inadequate laboratory diets and to the chronic diarrhea (and resulting water imbalance) caused by the enormous intakes of polysorbates. The Committee questioned "the validity of using levels above 10 percent in the assessment of the toxicological hazard of a food additive." Subsequent experiments on well-fed rats, hamsters, and dogs bore out the Committee's contentions. Levels of 5 percent and below were harmless, while greater amounts caused diarrhea and changes in some organs.

The polysorbates do not appear to be carcinogenic. Tumors did not develop when scientists fed polysorbate 60 to several species of rodents or injected it under the skin of baby mice. A four-generation study on rats conducted by a private testing laboratory showed that the emulsifiers do not affect reproduction. When polysorbate was injected under the skin of rats, some tumors did arise at the site of injection, but scientists do not believe that such tumors are indicative of a carcinogen.

The way the body digests polysorbates is well understood. The fatty acid portion (stearate, oleate) of the molecule is absorbed in the intestine and readily metabolized. The remainder of the molecule—polyoxyethylene sorbitan—is absorbed to the extent of 5 percent and subsequently excreted in the urine. The rest (95 percent) of the polyoxyethylene sorbitan is not absorbed and is eliminated in the stools.

Doctors have fed polysorbates to patients suffering from certain diseases to help them absorb fat in food into the bloodstream.

Ref.: FAO(17)-10, 12; FAO(29)-25; FAO(35)-127-145; CUFP-24; *J. Nutr. 61* 235 (1957); *J. Fd. Sci. 31* 253 (1966); *Tox. Appl. Pharm. 16* 321 (1970); 21 CFR 121.1008-9, 121.1030.

Potassium Bromate   GRAS

Millers and bakers have used potassium bromate since 1916 to artificially age and improve the baking properties of flour. Bromate is used at levels of 5 to 75 parts per million. Baking converts bromate (BrO_3^-) to bromide (Br^-), which is absorbed by the body when bread is digested. The bromide circulates harmlessly in the blood and is slowly excreted in the urine.

Calcium bromate is sometimes used in place of potassium bromate.

Ref.: FAO(35)-164.

Potassium Iodide (KI), Cuprous Iodide (CuI)   GRAS
Potassium or Calcium Iodate (KIO_3, $Ca(IO_3)_2$)

Iodine is one of the trace elements, small amounts of which are needed for man's proper growth, development, and health. In the early nineteenth century it was discovered that a deficiency of iodine in the diet causes goiter, the disease in which the thyroid gland grows to many times its normal size in a vain attempt to produce the iodine-containing hormone thyroxine. News of the miraculous effect of iodine spread like wildfire among sufferers of goiter, who quickly made iodine-rich seaweed a standard part of their diet. Ironically, seaweed contains so much iodine that while most people were cured, a few suffered from iodine-*excess* goiter.

In the 1910s and '20s, large-scale experiments proved that routine consumption of iodine could eradicate goiter in whole populations. Public health experts agreed that simply adding iodine to table salt would be the best, least expensive solution to an age-old problem. Salt manufacturers use 0.01 percent potassium iodide or cuprous iodide to supply the iodine.

Despite our thorough understanding of goiter and its virtual disappearance in the 1930s and '40s, this disease is now having a mild resurgence in the United States. Dr. Arnold E. Schaefer, of the Department of Health, Education and Welfare, recently supervised a national nutrition survey in low-income areas of the country in which the incidence of goiter was measured. On January 22, 1969, he reported to the Senate Select Committee on Nutrition and Human Needs that:

> Five percent of the total population examined to date exhibits an enlarged thyroid gland associated with low iodine intake. The World Health Organization classifies an area as having endemic goiter when five percent of the population has enlarged thyroid glands.

The reason for goiter in America is probably simple laziness: people assumed that goiter was "cured" and so they stopped reaching for iodized salt. Grocers, who usually display iodized salt and regular salt equally prominently, certainly have not done their part to educate the public. It must be recognized that goiter can never be cured in the way that malaria or the flu can be cured. There is no germ to destroy. Rather, there is a perpetual need for iodine that may be satisfied by naturally occurring iodine or by intentionally iodized foods. Persons living in inland regions should be sure to buy iodized salt.

For persons in many parts of the country, the major

source of iodine is not salt but bread. One of the important innovations that has enabled the baking industry to speed up production and keep down costs was the switch from a batch-type baking process to the continuous-mix technique. Along with the technological development came the need for a fast-acting dough improver (these chemicals make the dough drier and more manageable and the bread lighter). Potassium iodate has been the preferred chemical (usually mixed with the slower-acting bromates).

Bread usually contains 5 to 20 parts per million iodate, although the law permits up to 75 ppm. The legal limit amounts to approximately 2 milligrams per slice of bread. Most of the iodate (IO_3^-) is converted to iodide (I^-) during the preparation of the dough and baking of the bread. Any residual iodate is converted to iodide by the body.

The amount of iodine one ingests from iodate-treated bread is far greater than the amount—estimated to be 0.1 milligram per day—needed to prevent goiter. There is little chance, though, that the iodine in bread, added to that which is present in iodized salt and other foods, would cause iodine-excess goiter or otherwise prove harmful. Nevertheless, the FAO/WHO Expert Committee on Food Additives approached this matter cautiously and recommended in 1967 that iodates not be used as dough improvers. The Committee argued that:

> The use as a food additive for the treatment of a staple, such as flour, of a substance having such physiological significance and potency as iodate is highly undesirable. The Committee therefore recommended that iodates should not be used for flour treatment.

Azodicarbonamide, an approved food additive, is a suitable substitute for calcium or potassium iodate.

The iodate that quietly entered our food supply by way of the bakery initially confused the interpretation of radioactive iodine uptake tests used by endocrinologists to measure thyroid function. Once doctors realized that low iodine uptake was sometimes caused by iodated bread rather than thyroid malfunction, they were able to interpret their studies correctly. Hopefully, every endocrinologist is now aware of this possible problem.

Ref.: FAO(40)-15; FAO(40A)-112, 113; *Bull. World Health Org. 9* 211, 293 (1953); *New Eng. J. Med. 280* 1431 (1969); 21 CFR 121.1073.

Propyl Gallate   GRAS

Manufacturers add propyl gallate to foods to retard the spoilage of fats and oils. This antioxidant may increase slightly the shelf life of foods. Propyl gallate has not been adequately tested, frequently serves no useful purpose in foods, and should be avoided whenever possible.

Propyl gallate is used at levels up to 0.02 percent (of the fat or oil content) in animal fat, vegetable oil, meat products, potato sticks, and chicken soup base, and up to 0.1 percent in chewing gum. It is sometimes added to food packaging material, in which case it migrates as a vapor onto the food (breakfast cereals, potato flakes).

You will often find propyl gallate used in combination with BHT or BHA because of the synergistic effect these three food additives have in preventing fats from going rancid. The synergism permits a reduction of the total concentration of antioxidants in a food. Another reason for using propyl gallate in combination with other antioxidants is that high levels of propyl gallate produce an undesirable blue or green color in foods that contain minute amounts of iron or copper and moisture.

Rats—and presumably humans—have no trouble disposing of propyl gallate. The propyl ester is hydrolyzed to gallic acid, which is then converted in the liver to 4-methyl gallic acid. The latter compound is either directly excreted in the urine or coupled with another chemical (glucuronic acid) and then excreted. Metabolic studies of this type have not been, but should be, conducted in man.

Long-term studies on rats and mice have indicated that 0.2 percent propyl gallate in the diet has no effect on growth, internal organs, tumor incidence, or reproduction. Higher dosages affected growth and the kidneys. These feeding studies were conducted fifteen to twenty years ago and were not extensive or detailed enough to meet today's more rigorous standards. Because propyl gallate is so widely used, thorough long-term feeding studies should be conducted at once.

In an experiment reported in 1965, biologists found that high levels of propyl gallate damaged the cell division (mitotic) apparatus of liver cells. This observation underscores the need for detailed genetic and teratogenic studies.

As discussed in greater detail in the sections on vegetable oil and BHA and BHT, antioxidants are frequently added unnecessarily to oils and other foods. Because propyl gallate has not been adequately tested, the wise consumer can and should purchase brands of food that do not contain this or other synthetic antioxidants (BHA, BHT).

Ref.: FAO(31)-60; CRC-219, 704; C&EN-122; CUFP-11, 186; *Fd. Cos. Tox. 3* 457 (1965), *6* 25 (1968).

Propylene Glycol[41]   GRAS
Propylene Glycol Mono- and Diesters

Propylene glycol is one of several additives (glycerol and sorbitol are two others) that are used in foods to help

[41] Also called 1,2-propanediol.

maintain the desired moisture content and texture. Manufacturers add between 0.03 and 5 percent propylene glycol to candy, baked goods, icings, shredded coconut, and moist pet foods. This additive also serves as a carrier for oily flavorings and helps them dissolve in soft drinks and other water-based foods.

The human body has no trouble metabolizing propylene glycol; the body converts it to either chemical building blocks or to energy.

In a chronic feeding study done by FDA biologists, rats that ate food containing 4.9 percent propylene glycol suffered no ill effects. Shorter tests on dogs also indicated that this substance is safe.

Propylene glycol is closely related to glycerol and, like glycerol, can be chemically reacted with fatty acids. The propylene glycol mono- and diesters that form are analogous to mono- and diglycerides and natural fats and oils. Enzymes in the intestine convert the mono- and diesters back to propylene glycol and fatty acids, which are then absorbed and metabolized. Propylene glycol monostearate and other esters are used in shortening to help make lighter, fluffier cakes.

Ref.: FAO(35)-114; FAO(40A)-102; CRC-Ch. 11; CUFP-25, 268; *Arch. Biochem. 29* 231 (1950), *77* 428 (1958); *Poult. Sci. 48* 608 (1969); 21 CFR 121.1113 (propylene glycol mono- and diesters).

Quillaia   GRAS

Quillaia (pronounced kĭ-lié) is extracted from the soapbark tree and used to enhance the foam in root beer. This natural product has never been tested.

Quinine

In the early 1600s Jesuit priests living in Peru discovered that a substance in the bark of a local tree could cure malaria, the deadly, mosquito-borne disease. The miracle drug came to be called quinine. The sole natural source of quinine is the cinchona tree, a species indigenous to the mile-high eastern slopes of the Andes Mountains.

There is some evidence that quinine causes birth defects. Pregnant women should avoid quinine drugs and quinine-containing beverages.

Large amounts of quinine were shipped from South America to Europe from 1650 to 1850, during which time little attention was paid to the dwindling population of cinchona trees. To prevent the extinction of the species, and avert disaster, the Dutch began to cultivate the valuable tree in Java. A similar venture by the British in Madras, India, which state had the world's highest incidence of malaria, was a tragic failure. Fortunately cinchona flourished in Java and that island became the major supplier of the alkaloid. Quinine's heyday ended when Japan's capture of the Dutch East Indies in World War II forced the Allies to develop synthetic antimalarial drugs. These drugs have proved to be cheaper and more effective than quinine.

Aside from preventing the growth of the parasite that causes malaria, quinine has been used to induce abortions. It is also a nonspecific remedy for pain and fever (it reduces fever by dilating small blood vessels of the skin, thereby facilitating loss of body heat).

The sole use of quinine in foods is as a bitter flavoring in quinine water, tonic water, bitter lemon, and similar drinks. These beverages may contain quinine at levels up to 83 parts per million.

Reports in medical journals dating back to 1870 document quinine's ability to cause deafness in developing human and animal fetuses. A number of women who were either treated for malaria or who unsuccessfully attempted to abort their babies in the early months of pregnancy gave birth to deaf children. The quantities of quinine involved ranged from 600 milligrams to several grams. In a laboratory experiment, the auditory nerves of rabbit fetuses were damaged when the mothers consumed moderate dosages of quinine. This study was conducted in the 1930s and does not appear to have been followed up by further studies involving lower dosages and additional species. The sensitivity of the auditory apparatus to quinine is further indicated by the temporary hearing impairments that adults often experience when they are treated with quinine for malaria.

Tonic water and similar beverages contain approximately 83 milligrams of the drug quinine per quart. This level of drug in one or two gin and tonics a day would certainly not cause total deafness in a human fetus but could conceivably cause a temporary or permanent partial loss of hearing. Decreases of 1, 5, or 10 percent in hearing ability would never be spotted by pediatricians or parents, but such a side effect of a food additive would be intolerable. Until careful teratology studies are conducted, prudence would dictate that pregnant women not drink quinine-containing beverages.

Quinine water is definitely known to have at least one occasional deleterious effect. As few as five ounces of the beverage have caused purpura, a disease in which blood escapes from vessels near the skin, causing the victim to turn purple. More than fifty cases of this sensitivity to quinine have been reported.

The ability of quinine to cause cancer, mutations, or birth defects must be investigated before this additive can be considered safe.

Ref.: *Encyclopedia Brittanica* ("quinine," "cinchona," "malaria"); *Ann. Int. Med. 66* 583 (1967); *Pediatrica 32* 115 (1963); *Am. J. Obs. Gyn. 36* 241 (1938); 21 CFR 121.1081.

Saccharin

Saccharin is a synthetic sweetener that is three hundred to five hundred times sweeter than sugar and ten times sweeter than cyclamate. Because the body does not convert saccharin to glucose, persons suffering from diabetes use saccharin as a safe alternative to sugar. Saccharin can also help obese persons lose weight by supplying low calorie sweet-tooth satisfaction. Saccharin may be a weak carcinogen.

The way in which saccharin was discovered illustrates well the phenomenon of scientific serendipity (or, how sloppy, but intelligent, scientists make the most of their blunders). In 1879 Constantin Fahlberg, a chemist at Johns Hopkins University, ate a piece of bread that tasted unexpectedly sweet. Believing that he had accidentally contaminated his food with a chemical, Fahlberg returned to the laboratory and tasted some of the coal-tar derivatives that he had synthesized. Sure enough, one of them was incredibly sweet. The chemical was given the name saccharin and since about 1900 has been sold commercially. Incidentally, Fahlberg took out a patent for saccharin and made a good deal of money from his accidental discovery.

For half a century artificially sweetened foods were consumed almost exclusively by diabetics, but since the early 1960s manufacturers have sought to expand their market by encouraging the general public to buy artificially sweetened "diet" products. Most of the saccharin produced each year (4 million pounds in 1967) is used in carbonated beverages, dry beverage bases, and sweetener preparations (tablets, solutions) for home use. The remainder is used in dietetic foods (canned products, preserves, frozen desserts, etc.), pharmaceuticals, and a few nonfood applications (electrotyping). Soft drinks contain about 150 milligrams of saccharin per twelve-ounce serving; individual tablets or packets contain 20–50 milligrams (equivalent to one to two teaspoonsful of sugar). The warning that is usually printed in microscopic print on artificially sweetened products—"saccharin is a sweetener which should be used only by persons who must restrict their intake of ordinary sweets"—is not really taken seriously by manufacturers, the government, or the public. The advertising of saccharin-containing foods to the general public is a practice that should be investigated and stopped by the Federal Trade Commission.

The average person, as opposed to the strongest-willed dieter, does not automatically lose weight by eating artificially sweetened foods. The calories one avoids by drinking a bottle of diet soda are usually made up later in some other food. Although the casual dieter does not benefit from artificially sweetened foods, manufacturers have a good reason to continue to push their nonnutritive products: based on sweetening power, saccharin costs one twentieth as much as sugar. The low cost of saccharin, coupled with the premium prices that consumers often pay for diet foods, means that artificially sweetened products fatten profits, even if they have no effect on waistlines.[42]

The well-known gustatory shortcoming of saccharin, its disagreeably bitter aftertaste, enabled cyclamate, following its discovery in 1944, to capture a large share of the artificial sweetener market. Now that cyclamate has been

---

[42] One manufacturer of a saccharin product advertises that, "New ENZO-SWEET is 450 times sweeter than sugar . . . it sweetens your cost picture, too" (*Food Product Development*, February/March 1971).

banned from food because of its possible carcinogenicity, manufacturers devised ways of masking saccharin's aftertaste. Most commonly, they add about 0.2 percent glycine (aminoacetic acid), an amino acid, to the food.

The large amounts of sugar normally present in soft drinks, preserves, and other heavily sweetened foods add "body" or thickness to the foods as well as sweetness. When comparatively small amounts of saccharin replace sugar, the sweetness is there but not the body. Because of this, diet drinks and other products are frequently artificially thickened with vegetable gums, such as gum arabic or carboxymethylcellulose.

Now that saccharin is the sole artificial sweetener, record amounts of it will be consumed, but at the same time its safety is being closely scrutinized. As yet, there is only a slight suggestion of hazard.

Biochemical experiments have shown that saccharin is absorbed but not modified by the body. It is excreted unchanged in the urine.

The ability of saccharin to cause birth defects in mice, rabbits, and rats has been examined in well-designed experiments. Scientists fed saccharin to pregnant animals in the sensitive period of their pregnancy and did not detect any malformations in the fetuses. The trouble with these experiments was that the highest dose tested was only several times greater than that to which humans might be exposed. The negative finding would have been much more reassuring if dosages 25–50 times higher had been used.

A 1951 FDA study indicated that high concentrations of saccharin (above 0.1 percent of the diet) caused kidney damage in rats but did not induce cancer. The next good feeding study, this time in female mice, was reported in

1970 and, again, saccharin did not cause cancer. It should be added, though, that in both of these studies, as well as a third unpublished one, the researchers did not examine the animals' urinary bladders for the presence of cancerous tissue.

In 1957 scientists using an unusual technique obtained the first hint that saccharin might cause cancer. In this experiment the sweetener was not fed to mice but was mixed with cholesterol and surgically inserted in pellet form in the urinary bladders. Several mice developed bladder cancer. In the same study, massive amounts of saccharin injected into mice did not cause tumors. There is heated disagreement on how to interpret results obtained by this rather bizarre method. Many scientists contend that saccharin merely enhances the known tendency of cholesterol pellets to cause tumors. The scientists who did the experiment refrained from calling saccharin a carcinogen.

In March 1970—just after the great battle over cyclamates—University of Wisconsin biologists reported the results of an extensive study that involved the controversial bladder implantation technique. They found that the incidence of malignant tumors in the bladder was increased fourfold by saccharin: 50 percent of the mice exposed to saccharin developed tumors as compared to 12 percent of the control group. Again, because of the unusual route of administering the chemical, the carcinogenicity of saccharin was not proved. But the results of this experiment stimulated the Food and Drug Administration to review the safety of saccharin.

The FDA asked the Food Protection Committee of the National Academy of Sciences–National Research Council to review everything that was known about saccharin. In July 1970 the committee reported back that "the present and the projected usage of saccharin in the United States does not pose a hazard." But it did recommend that a variety of chronic toxicity, metabolic, and epidemiologic studies be conducted. By way of contrast, Dr. George Bryan, the Wisconsin scientist who conducted the study cited above, said "it may take many years before it is known exactly how dangerous the substance is and until then its use should be restricted to those who need it for medical reasons."

At the time of this writing, March 1972, several lifetime rodent feeding studies are nearing completion. One study, paid for by the sugar industry and conducted at the Wisconsin Alumni Research Foundation Institute, is scaring Monsanto and other manufacturers of saccharin. A preliminary analysis of the results indicated that a dietary level of 5 percent saccharin caused malignant tumors in the bladders of several male rats. If this report holds up and is supported by other studies, the FDA will be required by law to ban the general use of saccharin. Some saccharin and saccharin-containing products would still be produced, however, to serve the needs of those diabetics who must restrict their intake of carbohydrates.

"Damned if you do, damned if you don't" describes well the problem of choosing between sugar and saccharin. Sugar contributes to obesity and tooth decay; on the other hand, there is a small but distinct possibility that saccharin causes cancer. Neither sugar nor saccharin contains vitamins, minerals, or protein. Perhaps the time has come for sweet-toothed Americans to contemplate a third choice, which is the way people in many foreign lands eat: minimize the consumption of both sugar and saccharin by limiting the use of these chemicals in food and by abstaining from soft drinks, snack cakes, candy, and other highly sweetened manufactured foods.

Ref.: FAO(44A)-104; CRC-Ch. 14; ECT 19 593; Arzneimittelforschung 19 920, 925 (1960); New York Times, March 20, 1970; Fd. Cos. Tox 8 135 (1970); Science 168 1238 (1970); FDA review of the effects of saccharin on animals and man (January 26, 1970); "Safety of Saccharin for Use in Foods," NAS-NRC Report (July 1970); Washington Post, July 12, 1971; Chem. Eng. News, February 7, 1972, page 4; 21 CFR 121.4001.

Silicates[43]   GRAS
Silicon Dioxide

Everyone has been annoyed at one time or another by lumpy salt that just would not come out of the shaker. The problem arises when moisture adheres to grains of salt, enabling them to melt together; lumps form when hundreds of particles get into the act. Many persons put a few grains of rice into salt shakers to absorb the moisture. To save the consumer even that bit of effort, manufacturers have begun adding 2 percent silicates or silicon dioxide to salt. The additive coats the salt grains, keeping them from becoming damp and melting together. You will find silicate anticaking agents in powdered coffee whitener, vanilla powder, baking powder, dried egg yolk, and seasoning salts.

Silicon dioxide, or silica as it is often called, is the principal component of 95 percent of the earth's rocks. Combinations of silica with other minerals are termed silicates, also major constituents of the earth's crust. Thus, these additives are simply finely pulverized rock dust.

Silica occurs naturally at very low levels in all living organisms (10–200 mg/100g dry weight of human tissue). Silica in food is absorbed in the intestines to a lim-

---

[43] Aluminum calcium silicate; magnesium silicate; calcium silicate; sodium alumino silicate; sodium calcium alumino silicate; tricalcium silicate.

ited extent and then excreted in the urine and feces. Humans eliminate 10–30 milligrams of silica per day in the urine. It does not accumulate in our bodies.

Rats and rabbits have been fed diets containing 1 percent silica for ninety days. There was no effect on growth, survival, reproduction, blood, urine, and several internal organs. Humans have eaten 5–10 grams of silica a day for several weeks without apparent effect. The FAO/WHO Expert Committee on Food Additives has given silica and silicates a clean bill of health.

Ref.: FAO(46)-18; FAO(46A)-143; CUFP-269.

Sodium Benzoate   (benzoate of soda, benzoic acid) GRAS
Food manufacturers have used sodium benzoate as a preservative for over seventy years. Although it can prevent the growth of almost all microorganisms (bacteria, fungi, and yeast), it is effective only under acidic conditions. Thus, its use is limited to such foods as fruit juices, carbonated drinks, pickles, salad dressing, and preserves. It is used at levels of 0.05 to 0.1 percent. Sodium benzoate occurs naturally in many fruits and vegetables, notably cranberries (0.05 to 0.09 percent) and prunes, and so is no stranger to the human body. Moreover, in 1954 Dr. W. H. Stein reported in the *Journal of the American Chemical Society* that benzoate is a natural metabolite in our body.

This food additive has been tested in lifetime and short-term feeding experiments in man, dogs, and rats. In an experiment conducted in Germany, four generations of rats were continuously exposed to 0.5 or 1 percent sodium benzoate in their diet. Scientists did not observe any harmful effects on growth, life span, or internal organs; no tumors were detected. All the evidence indicates that sodium benzoate is quite safe, but careful experiments must be done to be sure that this additive does not cause birth defects.

In the United States our food may contain up to 0.1 percent sodium benzoate. Some nations permit as much as 1.25 percent benzoate in certain foods.

Ref.: FAO(31)-27; CRC-142-151; *J. Am. Chem. Soc.* 76 2848 (1954).

Sodium Carboxymethylcellulose (CMC;   GRAS
Cellulose Gum) and related compounds

Cellulose is a safe and inexpensive carbohydrate that comprises the woody parts and cell walls of plants. Chemists have learned how to modify cellulose to generate derivatives with a variety of properties. The nature and degree of modification affect the thickness of solutions, solubility, or concentration or temperature at which a solution gels.

Carboxymethylcellulose (CMC), the most widely used cellulose derivative, is made by reacting cellulose (wood pulp, cotton lint) with a derivative of acetic acid (the acid in vinegar). Manufacturers first used CMC as a food additive in 1924, but it was not used widely until the early 1940s, when natural vegetable products were in short supply due to the war. Some of the functions CMC serves in foods include:

—in ice cream: adds body and improves the texture, controls the formation of ice crystals; because CMC causes whey to separate out, it is often used in combination with carrageenan or gelatin (up to 0.3 percent);

—in beer: to stabilize the foam;

—in pie fillings and jellies: prevents fruit from settling or floating (up to 0.5 percent);

—in cake icings: prevents sugar from crystallizing and water from evaporating; helps maintain a smooth, glossy appearance;

—in diet foods: the water-binding capacity of CMC gives the dieter a feeling of fullness and satiety without adding any calories; adds body to artificially sweetened beverages (up to 3 percent);

—in bread doughs: helps hold in moisture, increases volume;

—in candy: prevents the sugar from crystallizing.

Animals cannot absorb or digest carboxymethylcellulose; when ingested it is totally excreted in the feces. Massive amounts (more than 5 percent) of CMC in food diminish an animal's rate of growth but have no apparent effect on internal organs, cancer incidence, or reproduction.

Tumors have never been induced in experimental animals by food containing as much as 20 percent CMC, but when researchers injected the additive under the skin of rats, tumors formed at the site of injection. The FAO/WHO Expert Committee on Food Additives and most scientists, however, agree that CMC is not carcinogenic, because it passes right through the gastrointestinal tract and never enters the blood or internal organs. Injection-site tumors are probably caused by physical irritation.

CMC has been used as a laxative and antacid. No undesirable side effects were caused by doses ranging as high as ten grams a day for six to eight months.

The food industry also uses the ethyl, methyl, methyl ethyl, hydroxypropyl, and hydroxypropyl methyl[44] derivatives for many of the same purposes as carboxymethylcellulose. These compounds are used to improve the clarity of berry pie fillings, retard melting of ice cream,

[44] The full names of these chemicals are ethyl cellulose, hydroxypropyl cellulose, etc.

reduce absorption of water by pie crust, substitute for a portion of the egg ingredient in cake batter, and artificially thicken many foods. An unusual feature of solutions containing methyl cellulose is that they become thicker rather than thinner upon heating; above a certain temperature they may gel. The body does not absorb or digest any of these substances.

Unmodified cellulose is sometimes used as an inert filler in solid and liquid diet foods, but ordinarily it gives foods an unacceptable grainy texture. Food technologists recently overcame this problem by treating cellulose with carefully controlled amounts of acid. They call the product of the reaction microcrystalline cellulose (MC) and use it as a food additive.

Microcrystalline cellulose's properties are interesting enough to suggest that it may become an increasingly familiar name on food labels. Particles of MC, which are extremely porous, can absorb such liquid foods as syrup, melted cheese, and liquified peanut butter and convert them to dry granular forms. Solutions of microcrystalline cellulose in water have a creamy consistency at low concentrations and form smooth spreading gels at high concentrations. According to the manufacturer of MC, unpublished animal feeding studies show that MC does not impair the availability of nutrients in the diet.

Ref.: FAO(40A)-75, 79, 82; FAO(46A)-131; CRC-Ch. 8; *Food Engineering* 33 44 (August 1961); *Fd. Cos. Tox.* 2 539 (1964), 6 449 (1968); GRAS (CMC, MC, methyl cellulose), 21 CFR 121.1021, 1087, 1112, 1160 (other derivatives).

Sodium Erythorbate[45]   GRAS

According to Chas. Pfizer & Co. sodium erythorbate assures "a more appetizing red 'showcase' color in proc-

---

[45] Sodium erythorbate is the sodium salt of erythorbic acid; it is also called sodium isoascorbate.

essed meats . . . When sprayed on presliced ham and bacon, sodium erythorbate retards color fading and preserves eye appeal." According to some consumers the use of erythorbate on meat is a deceptive practice because the additive helps make the meat look better than it really is.

Sodium erythorbate is a close but nonnutritive relative of vitamin C (ascorbic acid). Its most important use is to brighten the pink color of frankfurters, bologna, sliced pastrami, and other cured meats and to prevent that color from fading in retail display cases. Occasionally, it is used as an antioxidant in beverages, baked goods, pimento salad, and potato salad.

Erythorbate is not a vitamin, despite its structural similarity to vitamin C. But because oxygen reacts slightly more readily with erythorbate than with ascorbate, erythorbate helps protect the vitamin in certain foods.

Sodium erythorbate has not been extensively tested. In experiments on animals and man, biochemists found that this chemical does not replace vitamin C in the body and that it is fairly rapidly excreted, but only one inadequate chronic feeding study has been reported. That study, done in the early 1940s on rats by FDA scientists, did not reveal any problems. Thorough long-term feeding, genetic and teratogenic experiments must be conducted before this additive can be considered free from hazard. Meanwhile, meat packers should stop using erythorbate or switch to sodium ascorbate, which is certainly safe. At least one major packer, Oscar Mayer & Co., uses ascorbate.

Ref.: FAO(31)-23; *Food Tech. 12* (6) (1958), *19* 1719 (1965); *J. Nutr. 81* 163 (1963); Chas. Pfizer & Co. "Pfizer and Food" (1965); Ziegler, P. T., *The Meat We Eat,* Interstate Printers and Publishers, Decatur, Ill., 1966.

Sodium Nitrate and Sodium Nitrite (NaNO₃; NaNO₂)[46]

Except for cyclamates and MSG, no food additives have been more in the news than sodium nitrate and sodium nitrite. These additives are toxic at levels only moderately higher than the levels used in food, and accidental overdoses have caused many fatalities—most recently in March 1971. In fact, nitrite is one of the few food additives that is definitely known to have caused deaths in the United States. Nitrites and nitrates are also potentially dangerous because they can lead to the formation of cancer-causing chemicals. Despite their dangers, nitrite and nitrate are valuable additives under certain circumstances because they can prevent the growth of the bacterium (*Clostridium botulinum*) that causes botulism food poisoning. These additives are used to preserve ham, bacon, frankfurters, luncheon meats, and smoked fish. They also produce the characteristic pink color of cured foods and contribute slightly to their taste. The unusual properties of nitrite and nitrate challenge scientists to find a safer substitute or to devise ways of minimizing their hazards.

The discovery that nitrate and nitrite are useful in curing meat occurred quite by accident. Legend has it that salt-preserved meat, which is brown, frequently had isolated bright red patches, which resembled fresh meat. A little investigating revealed that the red patches were due to nitrate impurities in the salt. From then on merchants intentionally added nitrate to curing solutions to maintain or create an appearance of freshness in meat.

In 1899 scientists discovered that nitrate itself is not the active curing agent, but that it serves as a source of

---

[46] Potassium nitrate and nitrite are used interchangeably with sodium nitrate and nitrite.

nitrite. Shortly thereafter scientists learned that the nitrite decomposes to nitric oxide, which reacts with the brown, iron-containing pigments (heme) in muscle and blood. The reaction converts the pigments to stable, bright-red compounds, nitrosomyoglobin and nitrosohemoglobin.

The original reason manufacturers added nitrite to meat was to improve the color, but as a fortunate side effect nitrite can also prevent the growth of *C. botulinum*. Nitrite makes *botulinum* spores sensitive to heat. When foods are treated with nitrite and then heated, any *botulinum* spores that may be present are killed. In the absence of nitrite, spores can be inactivated only at temperatures that ruin the meat product. If spores are not killed, they may develop into bacteria and produce their deadly toxin. Nitrite's preservative action is particularly important in foods that are not cooked after they leave the factory, such as luncheon meat, smoked ham, and sausage. A preservative may also be necessary in thick foods, such as ham, because these offer an oxygen-free environment, the kind in which *botulinum* can grow. The toxin does not pose a danger in foods that are always well cooked, such as bacon, because the toxin would be destroyed in cooking. Laboratory studies demonstrate clearly that nitrite *can* kill *botulinum,* but whether it actually does in commercially processed meat is now being questioned. Frequently, the levels used may be too low to do anything but contribute to color.

While meat packers defend the use of nitrite primarily on the grounds that it kills bacteria, they relish it mainly because it makes meat redder. Fatty frankfurters are much more attractive and salable if they are pink rather than gray.

Government regulations allow cured food to contain up to 500 parts per million nitrate and 200 ppm nitrite. The only exception is smoke-cured tuna fish, in which the government allows only 10 ppm to enhance the color. In actual practice, according to analyses and surveys conducted by the American Meat Institute Foundation, meat contains up to 700 ppm nitrate and 480 ppm nitrite. FDA inspectors discovered a smoked salmon that contained a dangerously high level of over 3,000 ppm nitrite. Violations such as these of federal regulations reflect the traditionally lackadaisical enforcement attitudes of the Department of Agriculture and the FDA.

Nitrite is one of the most toxic chemicals in our food supply. Dozens of persons have died from nitrite poisoning and countless others have been incapacitated. Nitrite's toxicity is due to its ability to disable hemoglobin, the molecule in red blood cells that transports life-giving oxygen.

Nitrate and nitrite enter our food supply from intentional food additives, nitrate-contaminated water supplies, and vegetables grown in heavily nitrated (from fertilizer) fields. Most of the nitrate one ingests is excreted unchanged. Some fraction of it, however, may be converted to nitrite by intestinal bacteria. When absorbed into the bloodstream, nitrite readily converts the hemoglobin in blood cells to methemoglobin, which cannot carry oxygen. Human blood normally contains about 1 percent methemoglobin and 99 percent hemoglobin, and there is no problem. However, if nitrite is present and raises the percentage of methemoglobin to above 10 or 20 percent, the blood's ability to carry oxygen is severely impaired. This condition is known as methemoglobinemia. Victims quickly discolor and have difficulty breathing. Death may result when the methemoglobin level exceeds 70 percent.

Infants are much more susceptible to methemoglobinemia than adults. There are several interesting reasons for this difference in sensitivity, including the following:

—infant blood contains relatively little hemoglobin compared to adult blood;

—all hemoglobin is not the same: as much as 60 percent of infant hemoglobin is of a type that is more sensitive to nitrite than is adult hemoglobin;

—the fluid intake of infants is ten times that of adults (adjusted for body weight), so nitrate in the water supply presents special hazards to infants;

—an enzyme that converts methemoglobin back to hemoglobin is in short supply in infants;

—the less acidic environment in the stomach and small intestine of infants favors the multiplication of nitrite-producing bacteria.

All of these factors disappear within the first year of life.

Gerber, Beech-Nut, and Swift add nitrite and/or nitrate to some of their meat-containing baby foods. Gerber baby foods may contain up to 25 ppm nitrite, Beech-Nut products up to 10 ppm nitrite and 75 ppm nitrate. Heinz does not add either additive to their products.

In a letter to the author, Dr. Robert A. Stewart, Gerber's director of research, explained why his company used nitrite. According to Dr. Stewart:

> We should make it perfectly clear that there are no preservatives used in the preparation of baby foods, since they are sterilized by heat processing in hermetically sealed containers . . . To reiterate, nitrite and nitrate are not food preservatives. They are not used for the purpose of destroying or controlling bacteria. They do cure meat by producing chemical changes resulting in the characteristic color and flavor of cured ham, bacon, etc.

In other words, nitrite and nitrate are used in baby food solely for cosmetic purposes and do not benefit the infant consumer in any way whatsoever.

Most infant deaths due to methemoglobinemia in

America have occurred in rural areas, where babies were fed contaminated well water. No deaths due to the nitrite in baby food have been reported. Nevertheless, the FAO/WHO Expert Committee on Food Additives cautioned in 1965 that "it is impossible to make an estimate of an acceptable dose for babies of six months of age or less on the basis of animal experimentation or clinical experience. *Nitrate should on no account be added to baby foods . . . Food for babies should not contain added nitrite*" [italics added]. Gerber, Beech-Nut, and Swift: please note.

Adults most commonly fall victim to nitrite poisoning because of human carelessness. In one recent incident a jar that was supposed to contain "Spice of Life meat tenderizer" was accidentally filled with pure sodium nitrite. On March 14, 1971, a Washington, D.C., man sprinkled the deadly white powder on his dinner and died hours later. Ironically, this man died just a few miles from the headquarters of the FDA—whose job it is to guarantee pure foods—and just two days before a congressional committee began hearings on the use of nitrite in food. Other fatal mishaps have resulted from the substitution of nitrite for table salt in a restaurant's salt shakers and from the overcuring of homemade sausages.

One method of preventing fatal accidents would be to mix nitrite and nitrate with a harmless blue or black food coloring in such proportions that the coloring would be noticeable only when an excessive amount of the additive was present in a food. Surely some means must be devised to prevent these senseless deaths.

A second hazard associated with nitrite is that this highly reactive substance can combine with chemicals called secondary amines to form nitrosamines. This reaction could occur in growing plants, preserved food, or in the stomach. Nitrosamines are powerful carcinogens, teratogens, and poisons. In the past year government scientists have found significant levels of nitrosamines in cooked sausage, cured pork, dried beef, bacon, and fish.

Nitrite is added to more than seven *billion* pounds of meat and fish annually. Common sense dictates that we minimize the risks of nitrosamine formation and cancer by minimizing the use of nitrite and nitrate. These chemicals should be used only when there is a possibility of botulism. Purely cosmetic uses should be banned, as several countries have already done. Banning such uses would eliminate these additives from bacon, pet food, smoke-cured tuna fish, and baby food. In other foods, safer preservatives are often available. In the curing of fish, for instance, heat can substitute for nitrite. In fact, the laws of Canada and the State of Minnesota prohibit nitrite in fish.

In frankfurters and luncheon meats, it is unclear whether nitrite is necessary to prevent the growth of *botulinum.* In any case, several companies are currently producing preservative-free products without any reported problems.

The food in which nitrite is most likely to be necessary is ham, because ham provides the oxygen-free milieu in which *botulinum* grows. The Federal Government and the meat industry are currently doing experiments to determine if the small amounts of nitrite in ham really inhibit the growth of bacteria or serve only as color fixatives. If nitrite is needed to prevent bacterial growth, usage levels should be set by law to maximize consumer protection and minimize risks. If it is not needed, it should be banned immediately.

Ref.: FAO(38A)-31, 37; CRC-172-176; *Arzneimittel-Forschung* 13 320 (1963); *Br. Med. J. 1* 250 (1966); *Fd. Cos. Tox. 5-9* many articles (1967–1971); *Nature 225* 21, 302 (1970); pers. comm. from Beech-Nut, Gerber, Heinz, Swift companies; *Meat*

Science Review 5 (2) 1 (1971); *Washington Post*, March 20, 1971; Hearings, House Committee on Government Operations, March 1971; *Med. World News*, February 25, 1972, page 4; 9 CFR 318.7(c)(4), 21 CFR 121.1063-4.

Sorbic Acid   (calcium, sodium, or potassium sorbate)

GRAS

Food manufacturers have been using sorbic acid to prevent the growth of molds and fungi since the 1950s. You will find it added to cheese, syrup, jelly, cake, mayonnaise, soft drinks, wine, dried fruits, margarine, and canned frosting.

Sorbic acid is effective over a broad range of acidities (up to pH 6.5) and therefore more appropriate in many foods than sodium benzoate, which is effective only in acidic foods. Moreover, the human body treats this virtually odorless, tasteless additive as a food instead of as an unwanted foreign chemical. Sorbic acid cannot be used in foods that are pasteurized because it breaks down at high temperatures.

The sorbates are potent inhibitors of mold and fungi, but only marginally effective against bacteria. This specificity makes them perfect for use in cheeses, because the fermentative action of bacteria may proceed while mold growth is prevented.

Sorbic acid is chemically similar to fat (caproic acid, to be specific) and is metabolized by the body just as if it were a natural fat. Thus, our bodies use it as a food and source of energy. Sorbic acid can also be metabolized by microorganisms when there is a small quantity of sorbic acid and a large number of microbes. However, when a small number of microbes is confronted with a massive amount of sorbic acid—as is the case in treated foods—one or more of the microbes' enzymes is inhibited and growth stops. A few strains of mold are resistant to even high concentrations of sorbic acid, and these strains may contaminate treated foods.

Experiments on rats and dogs have shown that sorbic acid is safe. In a lifetime feeding study, German scientists found that rats suffered no ill effects when their food contained 5 percent sorbic acid. In fact, the male rats even enjoyed a 15 percent increase in their life span, which was possibly due to the chemical having prevented the growth of germs.

Ref.: FAO(40A)-60; FAO(31)-90; *ECT 18 589*; CRC-151; *Arzneimittel-forschung 10 997* (1962).

Sorbitan Monostearate

Sorbitan monostearate serves as an emulsifier in cakes, cake icing, whipped vegetable oil toppings, frozen pudding, coconut spread, and many other foods. It is used at levels up to 1 percent, often in combination with one of the polysorbate emulsifiers.

One of sorbitan monostearate's interesting applications is as a "bloom" inhibitor in chocolate[47] candies. "Bloom" is the discoloration which occurs on the surface of chocolate candies when they are warmed up and then cooled. The phenomenon is due to fats migrating to the surface in warm, melting candy. When the candy is cooled, the fat solidifies on the outside, obscuring the cocoa fibers which impart the normal dark color. Sorbitan monostearate prevents "bloom" by forming a thin layer on the cocoa fibers and impeding the flow of fat molecules.

Scientists have examined the biological effects of sorbitan monostearate and found that it is safe. In a chronic

---

[47] The cocoa fat in the coating of chocolate candy is generally replaced by another edible fat which melts at a higher temperature. The high melting point helps inhibit "bloom." However, because the fat does not melt well at body temperature, polysorbate 60, an emulsifier, is added to prevent the candy from tasting greasy.

feeding study conducted by FDA researchers, rats and dogs were apparently not harmed by a diet containing 5 percent sorbitan monostearate. In a study conducted by a private testing laboratory, four generations of rats ate food containing 20 percent sorbitan monostearate; their growth, reproduction, lactation, metabolism, behavior, mortality, and tissues appeared normal. Although this additive is composed of two chemicals (stearic acid and sorbitol) that the body can ordinarily use as sources of energy, it is less nutritious because the body absorbs only 50 to 75 percent of it from the intestine.

Persons suffering from certain diseases use sorbitan monostearate to help their bodies absorb fat from food.

Ref.: FAO(35)-107; *J. Nutr. 61* 235 (1957); *Tox. Appl. Pharm. 1* 315 (1959); CUFP-26; CRC-442; 21 CFR 121.1029.

Sorbitol   GRAS

Sorbitol is a close relative of the sugars and mannitol and, like them, has a pleasant sweet taste. It is about 60 percent as sweet as sugar. One manufacturer claims that "sorbitol is one of the 'miracle' products of modern chemistry—long known as a nutritive ingredient of many fruits, berries, and plants." If not exactly a "miracle," sorbitol does serve an interesting variety of purposes.

After we ingest it, sorbitol is absorbed into the bloodstream and converted to sugar, thereby providing calories. But because sorbitol is absorbed very slowly, blood sugar levels rise only slightly. This feature made sorbitol useful in the treatment of diabetics prior to the use of insulin. Foods sweetened with sorbitol instead of sugar provided diabetics with a relatively safe source of sweetness and energy and allowed them to decrease their intake of fats. These "sugarless" dietetic products contain as many calories and as much carbohydrate as sugar-sweetened foods and are useless for weight-watchers or persons who need low-carbohydrate diets.

Chewing gums and candy sweetened by sorbitol are relatively noncariogenic, because bacteria in the mouth metabolize sorbitol slowly or not at all. (Tooth decay is caused by acid produced by bacteria that grow in the mouth.)

Bottlers occasionally add small amounts of sorbitol to low-calorie drinks and foods to mask the bitter aftertaste of saccharin and to provide the "body" and "mouthfeel" normally provided by sugar.

Soft, sugar-based candy retains its "as made" firmness and chewing properties when it is "doctored" with 1 to 3 percent sorbitol; also shelf life is extended because the sorbitol inhibits the crystallization of sugar. In shredded coconut and other foods sorbitol acts as a sweetener and helps maintain the proper moisture levels.

Large amounts of sorbitol (1 ounce for children; 2 ounces for adults) have a laxative effect, but otherwise sorbitol is perfectly safe and wholesome. Sorbitol's safety is supported by a variety of studies on humans and animals.

Ref.: FAO(35)-96; CRC-Ch. 11; *Physiol. Rev. 42* 181 (1962); Atlas Chemical Ind. Bulletin "Atlas Products for the Food and Beverage Industry"; *Treatment of Diabetes Mellitis,* Joslin, E. P., Lea and Fabiger, Philadelphia, Pa. (1959).

Stannous Chloride (SnCl$_2$)

Food manufacturers sometimes use stannous chloride (tin chloride) as an antioxidant in soft drinks (15 ppm) and bottled asparagus (20 ppm). Stannous chloride reacts readily with oxygen, thereby preventing the oxygen from combining with chemicals in food and causing discoloration and offensive odors.

Food additives contribute relatively little to our daily

intake of tin. Toothpaste containing stannous fluoride and canned goods are more important sources.

Studies on the biological effects of tin are progressing from two directions. Toxicologists are seeking to determine whether chronic ingestion of tin compounds has any adverse effects. They have found that stannous chloride and other tin compounds are poorly absorbed by the body and do not accumulate to any great extent. Thus, tin appears to be harmless.

On the other hand, nutritionists are interested in learning whether tin is a necessary ingredient in our diet. In an experiment reported in 1970, one group of rats ate food which contained all the known essential nutrients and trace elements, but from which tin was rigidly excluded. Another group of rats ate the same food plus a dose of tin. The rats whose diet contained tin grew significantly faster than those deprived of tin. This result means that small amounts of tin may be essential for normal growth.

Ref.: FAO(43)-16; *Bio. Bio. Res. Comm. 40* 22 (1970); *Fd. Cos. Tox. 3* 271, 277 (1965); *J. Nutr. 96* 37 (1968); 21 CFR 121.1213.

Starch, Modified Starch   GRAS

Starch is the major component of flour, potatoes, and corn and is normally thought of as a food, not a food additive. However, because starch can absorb large amounts of water, manufacturers and individuals use it to thicken homemade or commercially prepared soup, gravy, and other foods.

Natural starch is an effective, wholesome, and inexpensive thickening agent, but it has several technical limitations. It does not dissolve in cold water, solutions are not stable in the presence of acids, and it separates from water on standing. Food technologists have found that these and other limitations can be overcome by treating starch (usually cornstarch, less often potato starch) with any of a variety of chemicals.

Treating starch lightly with an oxidizing agent (oxygen, chlorine, potassium permanganate) bleaches the colored impurities (xanthophylls), but does not affect any other physical properties. Manufacturers add the ultra-white product to confectioners sugar and baking powder to absorb moisture and prevent caking.

Amylose

Amylopectin

Treating starch with acid (hydrochloric or sulfuric) or alkali (sodium hydroxide) breaks the large starch molecules into smaller pieces. The pieces, called dextrin, are a normal component of the diet and perfectly safe. Dextrin solutions are much thinner than solutions containing an equal weight of untreated starch. (See separate discussion of dextrin.)

Other kinds of modification make relatively great changes in the physical and chemical properties of starch. To understand these modifications and their effects, we must first examine the chemical structure of starch. Starch molecules are long, thin chains composed of hundreds or thousands of glucose (a sugar) molecules linked end to end like the cars in a long train. Linear chains are called amylose, branched chains amylopectin.

In one general class of modified starches, chemicals are used to tie chains together (whether the chains are linear or branched is irrelevant). These are called cross-linked starches.

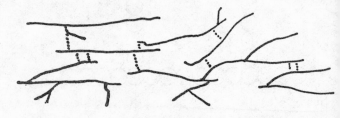

Cross-linked starch

A very small number of bridges (one per 500 to 1000 glucose units) stabilizes starch solutions against the effects of acids and strong agitation. Cross-linking increases the thickening ability of starch, thereby decreasing the amount needed in a food. In addition, cooking has relatively little effect on the thickness of solutions of cross-linked starch. This combination of properties makes cross-linked starch the ideal thickening agent for fruit pies. The major limitation of cross-linked starch is that it comes out of solution when a product is frozen and thawed.

Starch can also be modified by the addition of side chains. A representation of a molecule of "derivatized starch" is shown in the figure below:

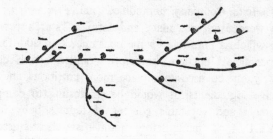

Derivatized starch

Side chains that carry a negative charge, such as acetate, succinate, or phosphate, cause starch molecules to repel one another. The repulsion prevents molecules from clumping together or crystallizing, which normally causes starch to separate out of solution or to gel. Solutions of derivatized starch have increased viscosity and clarity. You are most likely to encounter derivatized starch in frozen foods, in which they serve as thickening agents.

Starch may be modified by more than one chemical at a time. For instance, it may be cross-linked with one agent and then reacted with a second one that adds side chains. Cross-linked and derivatized starches possess most of the attributes of the singly modified substances. Thus, solutions are not affected by freezing and thawing or acids, and products in which they are used have long shelf lives. You will find these starch derivatives in foods in which a smooth, nongrainy, and noncohesive texture is desirable, such as baby food and fruit- and cream-pie fillings.

Manufacturers add derivatized starch to bottled baby foods to improve their consistency and to keep the solids in suspension. According to a letter to the author from H. J. Heinz Company, unmodified (native) starch would, if it were used, coagulate and form a "starch product plug which is unusable by the consumer." In addition to its beneficial function, however, modified starch could conceivably be used to replace more nutritious and expensive ingredients. It would be interesting to compare the meat and vegetable content of today's baby foods with baby foods made before modified starch was used.

Because modified starches are used in many infant and regular foods, it is of utmost importance that they be safe and nutritious. Modifications could affect the wholesomeness of starch either by decreasing its digestibility or by generating a toxic substance.

Cross-linking agents affect only a minute fraction of dextrose units in starch, and biochemists have found that our bodies can digest and utilize these starches. On the other hand, agents that add side chains affect a significant fraction of dextrose units. Derivatized starches, such as the starch acetate that is used in baby food, may be digested less completely than untreated starch.

Any treatment that adds side chains to or cross-links

starch molecules could introduce a hazard, and, therefore, modified starches should be evaluated in scientific studies. Unfortunately, most modified starches have been studied only in brief ninety-day rat experiments, if at all. No lifetime feeding studies have been conducted. The FAO/WHO Expert Committee on Food Additives noted in its 1970 report that modified starches are widely used but inadequately tested and urged that considerable additional research be conducted.

Ref.: FAO(46); (46A); ECT 18 685; CRC-Ch. 9; NAS-NRC Report, "Safety and Suitability of Modified Starches for Use in Baby Foods" (September 1970); Whistler, R. L., and Paschall, E. F., Starch: Chemistry and Technology, Vol. 2, Academic Press, New York (1967); 21 CFR 121.1031 (modified starches).

Stearyl Citrate   GRAS
Isopropyl Citrate

Manufacturers use citric acid as an antioxidant, but because it does not dissolve in fats and oils it can be used only in water-based products. This limitation was overcome when some clever chemist thought to react citric acid with either isopropyl alcohol or stearyl alcohol, both of which dissolve readily in oils and enable citrate to dissolve in oil. The compounds that form, stearyl citrate and isopropyl citrate, protect oils by trapping metal ions that might otherwise catalyze oxidative reactions and cause rancidity.

Stearyl citrate is used in margarine (up to 0.15 percent). Isopropyl citrate is used in vegetable oil and other fat-containing foods at levels up to 0.02 percent.

These two food additives were last tested on animals over twenty years ago. Lifetime feeding studies, which encompassed four generations of rats, showed that diets containing up to 10 percent of either additive did not affect growth, tumor incidence, health, or reproduction. Biochemists have found that the body converts these additives to citrate and stearyl or isopropyl alcohol, all of which are digestible and harmless.

The FAO/WHO Expert Committee on Food Additives recommended in 1969 that further feeding tests be conducted, with special attention given to the effects of these substances on the liver and kidney.

As discussed in the sections on vegetable oil and butylated hydroxyanisole (BHA), antioxidants such as these are often added unnecessarily to foods; you can avoid them by reading food labels carefully and purchasing additive-free brands.

Ref.: FAO(31)-51; FAO(40A)-54; FAO(46A)-112; *Food Res. 16* 258, 294 (1951).

Succistearin

Succistearin is a new emulsifier that is used to a limited extent in shortening to help make more tender baked goods. The full, mind-boggling name of this food additive is stearoyl propylene glycol hydrogen succinate.

Biochemists have shown that succistearin is metabolized well by rats (and presumably humans). The body converts the additive to succinic acid, stearic acid, and propylene glycol, all of which may be used as sources of energy.

Ref.: Tox. Appl. Pharm. 17 519 (1970); 21 CFR 121.1197.

Sucrose   GRAS

Sucrose, ordinary table sugar, is the most widely distributed sugar and is present in all parts of all plants. Sugar cane, the tropical plant that supplies two thirds of our sugar, grows in many parts of the world, but it originated in northeastern India. Traders carried it westward to Egypt and eastward to China about fifteen hundred years ago. A thousand years later, in 1494, Christopher Columbus brought the valuable plant to the West Indies. Other sources of sugar are sugar beets, maple trees, honey, sorghum, pineapple and other ripe fruits.

Americans consume enormous amounts of sugar . . . approximately one hundred pounds per person per year. The sugar we get from soft drinks, cakes and cookies, preserves, presweetened breakfast cereals, desserts, and the sugar bowl contributes greatly to this nation's incredible toll of tooth decay, obesity, and heart disease. It may even cause a form of cancer. Dr. Dennis Burkitt, an English cancer specialist, has postulated that a diet high in refined carbohydrates leads to high bacterial count and slow bowel movements, which in turn enhance the likelihood of cancer of the colon.

Pure, white, granulated sugar, unfortunately, is as pure as its makers claim; the refining process removes every last trace of vitamins, minerals, and protein. Sugar that has been extracted from cane but not yet refined is called raw sugar. It contains small amounts of trace minerals in addition to the sucrose, but it still makes you fat and rots your teeth. As a salesman in a health food store said, "Raw sugar and refined sugar are both rotten foods, but at least raw sugar contains some minerals."

There is no denying that sugar and sweetened foods taste good and supply energy. However, Americans have gone overboard in their affection for sugar. Except for those who enjoy large dentist bills and large girths, most of us should decrease our consumption of sugar.

Ref.: Fd. Cos. Tox. 9 439 (1971); *Cancer 28* 3 (1971); Senate Consumer Subcommittee, *Hearings* of March 2, 1972.

Sulfur Dioxide (SO$_2$)   GRA$

Sodium Bisulfite (NaHSO$_3$)[48]

Food processors treat foods and beverages with sulfur dioxide or bisulfite to prevent discoloration and to inhibit the growth of bacteria. Historians have traced the use of these chemicals back thousands of years to when the Romans and Ancient Egyptians used them as preservatives in wine.

Bisulfite (a powder) and sulfur dioxide (a gas) protect carbonated drinks, wine, grape juice, grapes, sliced fruits and vegetables, dehydrated potatoes, powdered soup mixes, etc. Maraschino cherries are bleached with this additive before they are dyed red or green. Sulfur dioxide has a strong, pungent odor which would warn consumers of foods containing excessive amounts (above 0.05 percent).

Bisulfite is a very reactive chemical, a property which defines its uses and its limitations. Bisulfite prevents foods from discoloring by combining with sugars[49] and enzymes in the food. In the absence of bisulfite, sugars tend to react with other chemicals to form colored compounds, while enzymes catalyze reactions which generate colored compounds.

On the negative side of the ledger, bisulfite destroys vitamin B-1 (thiamine) and therefore is banned by law from foods rich in this vitamin. Bisulfite can also restore a "fresh-red" color to old or spoiled meat; this deceptive use is specifically prohibited by law. In 1968 and again

---

[48] And related compounds: sodium metabisulfite Na$_2$S$_2$O$_5$; sodium sulfite, Na$_2$SO$_3$; potassium equivalents of these salts. When sulfur dioxide or any of these salts are dissolved in water, bisulfite forms.

[49] Aldehydes and ketones, also.

in 1969 FDA inspectors caught the Tender Treat Company of Spokane, Washington, marketing "B. T. Seasoning." This product contained sodium sulfite and sodium bisulfite and was intended for use on meat—a flagrant violation of the law. The merchandise was condemned, but no other penalties were invoked.

Animals, including humans, convert bisulfite to sulfate, which is innocuous and excreted in the urine. In a lifetime feeding experiment in which three generations of rats drank water spiked with 0.35 or 0.75 percent bisulfite, no harmful effects were seen. Other experiments showed that dietary bisulfite did not affect the vitamin A level in the liver (sulfur dioxide can destroy that vitamin). Adverse effects that are occasionally seen in feeding studies are usually due to malnutrition caused by the destruction of vitamin B-1 by bisulfite.

Under acidic conditions this additive causes mutations in bacteria, but it is doubtful that this observation has any relevance for mammals. Genetic studies have not been carried out on rodents.

Ref.: FAO(40A)-63; *J.A.C.S. 57* 536 (1935); *J. Pharm. Pharmacol. 12* 488 (1960); *Sulfur Dioxide,* L. Schroeter, Pergamon, New York (1966); *Nature 227* 1047 (1970); CUFP-11; *ECT 19* 419; *FDA Papers,* July/August 1969, May 1970.

Tannin, Tannic Acid   GRAS

Tannin, or tannic acid as it is sometimes called, is a mixture of chemicals obtained from the bark, leaves, and galls of a wide variety of shrubs and trees. We ingest significant amounts of tannin (100 to 500 milligrams) with every cup of tea, coffee, and cocoa that we drink. The name tannin is derived from the traditional and major use of this substance, the tanning of leather.

Food manufacturers take advantage of tannin's taste

and chemical properties. They use tannin as an ingredient in butter, caramel, fruit, brandy, maple, and nut artificial flavorings. Manufacturers employ tannin's ability to form insoluble complexes with proteins and other substances to remove undesired material from beer, wine, and oils.

In the 1930s doctors used tannin as a burn ointment. They sprayed it over the burnt area, where it formed an artificial scab that alleviated the pain and enhanced the healing process. This treatment was discontinued when physicians discovered that tannin sometimes entered the bloodstream and damaged the liver.

Although tannin is on FDA's GRAS ("Generally Recognized As Safe") list of food additives, scientists suspect that it might be a weak carcinogen. Their suspicion was aroused by experiments in which tannin was injected under the skin of rats and mice, causing tumors to develop both at the site of injection and in the liver. The injection-site tumors are not significant, but the liver tumors justify concern. Tannin must be subjected to a complete battery of studies before we can consider it safe.

Ref.: FAO(40)-13; CUFP-193, 270; Cancer Res. 19 501 (1959); Prog. Exp. Tumor Res. 2 245 (1961); Fd. Cos. Tox. 7 364 (1969).

Tartaric Acid   GRAS

Tartaric acid occurs naturally in grapes and other fruits and is made commercially from waste products of wine production. It was once more widely used than it is today, but American industry began to replace tartaric acid with other acids, especially fumaric, when World War II wreaked havoc with its price and supply. Tartaric acid, as one might guess from its name, has an extremely tart taste.

Tartaric acid is a constituent of grape and other artificial flavors that are used in beverages, candy, ice cream, baked goods, yogurt, and gelatin desserts. It also serves as the acid in some baking powders.

Biochemists have found that most of the tartaric acid we ingest is destroyed in the intestines by bacteria. The fraction (20 percent) that is absorbed into the bloodstream is rapidly excreted in the urine. Guinea pigs and humans metabolize this additive identically, but rabbits, dogs, and rats go to the trouble of absorbing, then excreting, the entire dose.

People have consumed large amounts of tartaric acid as a laxative without apparent harm. In 1947 Drs. Fitzhugh and Nelson, FDA biologists, published the results of lifetime feeding studies on rats; they concluded that dietary levels as high as 1.2 percent were harmless.

Ref.: FAO(31)-96; CRC-261-3; CUFP-17, 46, 193.

Textured Vegetable Protein (TVP)
Isolated Soy Protein

Isolated soy protein and textured vegetable protein are being used in a rapidly growing number of manufactured foods. Isolated soy protein is simply protein purified from soybeans. Textured vegetable protein is soy protein that has been combined with chemical additives and processed into granules, chunks, or strips that resemble meat.

Soy protein was originally developed to help solve the protein shortage and hunger problems in underdeveloped nations, but its versatility and low cost are making imitation meat products increasingly popular in the United States. The composition of products made from soy protein can be selected arbitrarily by the manufacturer. A typical dehydrated item may contain 35 to 50 percent protein, 20 to 30 percent carbohydrate, and 1 to 20 percent fat. The natural flavor of soy-based products is bland and nut-like, but seasonings and flavorings can make it taste remarkably like chicken, pork, ham, bacon, pepperoni, fruit, nut, or other flavors. Soy-based imitation meats are a godsend to persons who cannot eat meat for religious reasons.

The major ways in which we will be consuming textured soy protein in the coming years were described in one company's promotional brochure. We are advised that:

> TVP enhances nourishment, so it is well suited for institutional feeding, restaurants, and drive-ins . . . It is an excellent protein ingredient for casseroles, patties, meat balls, pizza toppings, sandwich fillings, stews, salads, snacks, and convenience foods. TVP is especially applicable to dietary foods whose vegetable origin and control of fat content are of prime importance.

It may be added that initially restaurants, hospitals, and other institutions will be the big market for imitation meat products because they do not have to tell their patrons the ingredients of the food.

Products based on soy protein are undeniably nutritious, because the biological value of soy protein is similar to that of meat protein. However, imitation meat products may lack the vitamins and minerals contained in the real thing. Most meats contain good amounts of thiamine (vitamin B-1), riboflavin (B-2), niacin, iron, and phosphorus; liver and kidney contain vitamin A and the B vitamins, trace minerals and iron and phosphorus. The FDA plans to specify the amounts of nutrients that will have to be added to imitation meat products.

Imitation meat is seasoned, held together, and colored with a flock of flavor enhancers, thickening agents, emulsifiers, artificial colorings and flavorings. A recipe for artificial bacon includes soy protein, hydrolyzed yeast protein, salt, spices, monosodium glutamate, vegetable gum stabilizer, yeast, corn oil, artificial coloring, and water. The wholesomeness of the food depends on the nature of each of the additives. Obviously, these new products do not relieve shoppers of the eye-straining job of scrutinizing labels.

Ref.: Protein Food Supplements, Noyes, Robert, Noyes Development Corp., Park Ridge, N.J.; Archer-Daniels-Midland Co. Brochure "TVP/A fabulous new food"; *Wall Street Journal,* October 2, 1969.

Thiodipropionic Acid   GRAS
Dilauryl Thiodipropionate

Thiodipropionic acid and its dilauryl derivative are occasionally used in foods and food packaging for the same purpose as BHA, BHT, and propyl gallate: to prevent fats and oils from going rancid. These compounds function by reacting with oxygen before the oxygen can react with fat. They may be used at concentrations up to 0.02 percent.

The most recent chronic feeding studies of thiodipropionic acid and its dilauryl relative are over twenty years old. The studies, conducted by FDA toxicologists, indicated that the two additives are safe at levels up to 1 percent of the diet, but the scope of the studies falls far short of today's standards. Because these substances are neither widely used nor irreplaceable, the FDA should ban them from foods until tests show that they do not cause cancer, birth defects, or other harm.

Ref.: FAO(31)-99; *Adv. Fd. Res. 3* 197 (1951); CUFP-12.

Tragacanth Gum   GRAS

Gum tragacanth exudes in the form of ribbons from the damaged roots of a small bush that grows wild in Iran. U.S. firms imported 1.7 million pounds of this vegetable gum stabilizer in 1970.

Tragacanth exhibits a resistance to acids which is unexcelled among vegetable gums. This property makes it the ideal thickening agent for acidic foods, such as vinegar-containing salad dressings.

No long-term feeding or biochemical studies have been conducted, so it is impossible to evaluate the safety of this additive.

Ref.: FAO(46A)-104.

Vanillin, Ethyl Vanillin GRAS
Vanilla

In 1520 the Aztec emperor Montezuma gave Cortés, the Spanish explorer, an unbelievably delicious drink of cocoa. It did not take the curious European long to discover that the drink's great taste was due in part to the addition of extracts of a local climbing orchid, the vanilla plant.

Cortés brought some vanilla back to spice-starved Spain, where it caused a minor sensation. Within a few years the Spaniards built factories to manufacture chocolate flavored with vanilla in their native land.

The vine-like vanilla plant is indigenous to Mexico, Central America, northern South America, and a few other scattered locations. Early attempts to cultivate vanilla plants in regions more convenient to the European market were unsuccessful; the plants would sometimes grow and flower, but they would never produce their valuable fruit. The reason for this was discovered in the 1800s: the flowers of the plant are structured in such a way that they cannot self-pollinate. In regions where vanilla was indigenous, unique varieties of hummingbirds and bees pollinated the flowers; these varieties were found nowhere else in the world. This is a classic case of interdependent species evolving in tune with one another.

Once the pollination problem was understood, plants

were pollinated by hand, and vanilla farms sprang up in several corners of the world. The Malagasy Republic (Madagascar) is currently the world's leading vanilla producer, accounting for 75 percent of all vanilla production. The United States consumes half of the world's vanilla.

Vanilla is probably the most widely used food flavoring. Manufacturers add it to ice cream, baked goods, beverages, chocolate products, gelatin desserts, and candy at concentrations ranging from 0.006 to 0.1 percent. The vanilla bean is an ingredient in many flavors—butter, root beer, fruit, chocolate, and, of course, vanilla.

The great demand for vanilla far outstripped the rather meager (and expensive) natural supplies years ago. The chemical industry responded to this situation by manufacturing synthetic vanillin, the major flavoring component of vanilla. Vanillin has a similar taste but is not quite as good as the real thing, because natural vanilla contains a multitude of minor constituents that modify and perfect the flavor.

Flavor chemists have synthesized a variety of derivatives of vanillin in an attempt to reproduce the exact taste of vanilla in a single chemical. So far these efforts have failed, although one of the modified molecules, ethyl vanillin, is valuable, because it comes closest to matching the desired taste and because it has 3.5 times the flavoring power of vanillin.

The metabolism of vanillin is well known. The body absorbs it, converts most of it to vanillic acid, and then excretes both compounds in the urine.[50] The metabolism of ethyl vanillin has not been studied, but the body probably handles it similarly to vanillin.

[50] A large fraction of vanillin and vanillic acid are combined with sulfate or glucuronate prior to excretion.

Lifetime feeding tests on rats of vanillin and ethyl vanillin did not reveal any adverse effects, but so few animals were used that the experiments must be repeated on a larger scale.

The harmlessness that was indicated for vanillin by feeding and metabolic studies and hundreds of years of usage was further supported by the discovery that two hormones, adrenalin and noradrenaline, are degraded to vanillin when they are excreted by the body. The body's own production of vanillin is as much as a half milligram per day.

Because ethyl vanillin is not a natural metabolite, it should be thoroughly tested. Its metabolism in man and its effect on animal reproduction must be investigated before we can consider it safe.

Ref.: FAO(44A)-39, 78; *Econ. Bot.* 7 291 (1953); CRC-472, 746; CUFP-123, 202.

Vegetable Oil

Soybeans, peanuts, cottonseeds, and similar plant products contain large amounts of natural oils, which can be extracted and used as a food or food ingredient. The low cost and great versatility of vegetable oils, as compared to butterfat, have stimulated the development of a myriad of products in which they play a key role, such as in imitation milk, powdered nondairy creamers, imitation whipped cream, margarine (imitation butter), and cooking oil. These products are useful to the general population and a boon to persons whose religion or health require abstinence from dairy products or animal fat.

A vegetable oil molecule consists of two parts, a glycerol backbone which is not too interesting, and the fatty acids which are very interesting. The body needs—but cannot itself synthesize—several fatty acids, so it is vital that we eat foods that contain them. The most important of these essential fatty acids, linoleic acid, occurs in especially high concentration in safflower and corn oils (see Table III below).

Another important feature of fatty acids is that some are saturated, some unsaturated, and others polyunsaturated. The chemical difference between saturated and polyunsaturated oils is based on the number of hydrogen atoms in a fat molecule. A more important difference (for us) is that scientists have found that persons who eat large amounts of saturated fats generally have high cholesterol levels in their blood. These high levels, in turn, are frequently associated with "hardening of the arteries" (atherosclerosis), coronary heart disease, and heart attacks. The precise interrelationship between saturated oils, cholesterol, and hardening of the arteries is the subject of intense medical investigation, because heart and other arterial diseases kill more than 800,000 Americans every year.

Most doctors agree that eating excessive amounts of saturated fats contributes to hardening of the arteries, and, therefore, they advise their patients to moderate their intake of such fats. The American Heart Association's advice is clear and to the point:

Put LESS emphasis on foods high in saturated fats.
Put MORE emphasis on foods high in polyunsaturated fats.[51]

The table below rates a variety of vegetable and animal fats and oils according to their content of unsaturated and polyunsaturated fatty acids and their content

[51] The American Heart Association also recommends that we restrict our intake of cholesterol-containing foods, such as milk and eggs.

of linoleic acid; in each case the higher the number the better. Note that coconut oil, which is commonly used

Table III: Comparison of Fats and Oils*

Fat or oil	% linoleic acid	Relative amount of unsaturated plus polyunsaturated fatty acids
coconut	2	9
butterfat	2	30
cocoa butter	2	40
lard	14	73
olive	15	85
high oleic safflower	10	93
peanut	31	98
rice bran	32	110
sesame (USA)	43	110
cottonseed	54	110
corn	57	125
soybean	50	130
sunflower	68	130
safflower	75	132

* Adapted from *Food Engineering* (May 1970), courtesy of Drew Chemical Corp., Parsippany, N.J.

in imitation dairy products, is more highly saturated than butterfat or lard and, therefore, cannot help lower cholesterol levels. Manufacturers routinely list coconut oil on food labels as "vegetable oil," taking advantage of the fact that most persons mistakenly assume that all vegetable oils are equally wholesome. The FDA could outlaw this deception, which is a real health hazard, by demanding that oils be listed as cottonseed oil, coconut oil, etc., instead of "vegetable oil." The FDA should also require products containing coconut oil to bear an appropriate warning to the consumer, such as *"should not be used by heart patients."*

Vegetable oils are used in margarine on account of their low cost. However, because oils are liquid at room temperature, they must be modified before they are suitable for such use. The modification, which entails a reaction with hydrogen, converts some of the polyunsaturated oils to mono-unsaturated or saturated oils. This process is known as "hardening," "saturation," or "hydrogenation." Partially hardened or hydrogenated oils are semisolid at room temperature and thus ideal for use in margarine. Unfortunately, partially hardened oils are somewhat less nutritious than unmodified oils. During hydrogenation a substantial fraction of the linoleic acid is converted to a less valuable substance, oleic acid. Moreover, partially hardened oils are lower in polyunsaturates than the original oils. In recent years, manufacturers of margarine and shortening have been using improved methods, which result in smaller losses of polyunsaturated oils.

Polyunsaturated oils can react with the oxygen in air and develop a rancid odor and taste, especially in the presence of sunlight or metal ions. In living plants and animals, naturally occurring tocopherols (vitamin E) and other antioxidants prevent the oxidation. The antioxidant content of oils is reduced only slightly during commercial processing and should be sufficient to prevent the oils on supermarket and kitchen shelves from going rancid for a reasonably long period of time. Despite the natural content of antioxidants, some manufacturers add such synthetic antioxidants as BHT, BHA, or propyl gallate to their products. Whether the antioxidants are added because the oils are deficient in tocopherols, out of habit, because of poor production techniques, or in order to increase slightly the shelf life is not known. What is known, though, is that competing manufacturers produce perfectly acceptable products without adding antioxidants. As discussed in separate entries (see "butylated hydroxyanisole and butylated hydroxytoluene," "propyl gallate"), many of the synthetic antioxidants have not been adequately tested and may accumulate in the human body. The reader should not accept the unnecessary use of questionable food additives and should purchase only those brands of oils and shortenings that do not contain preservatives. Table IV lists the preservatives in some popular brands of oils and shortenings.

Table IV   Preservatives in Oils and Shortenings

Oil or shortening	Preservative
Wesson Oil (soybean, cottonseed)	none
Safeway "Nu-made" Corn Oil	none
Planter's Peanut Oil	none
Wesson Buttery Flavor Soy Oil	none
Crisco solid shortening (made from vegetable oils)	none
olive oils	none
Safeway "Nu-made" Oil (soybean, cottonseed)	BHT, BHA
Crisco Oil (soybean)	BHT, BHA
Safeway "Empress" Safflower Oil	propyl gallate, citric acid
Mazola Corn Oil	isopropyl citrate
Spry solid shortening (made from vegetable oil)	BHT, BHA
Safeway "Velkay" solid shortening (made from animal fat and vegetable oils)	"oxygen interceptor" (presumably BHT and BHA)
Kraft "Pure" Safflower Oil	BHT, BHA, citric acid

Ref.: American Heart Association; Harris, R. S., and von Loesecke, H., *Nutritional Evaluation of Food Processing*, John Wiley, New York, 1960.

Yellow Prussiate of Soda

Some manufacturers add yellow prussiate of soda (sodium ferrocyanide) to salt when they crystallize it. The additive generates jagged and bulky crystals, which resist caking; this mitigates the need for extra anticaking agents.

Although this additive contains cyanide, it is not toxic because the cyanide is tightly bound to iron atoms. Individuals have attempted to commit suicide by swallowing teaspoonfuls of this dangerous-sounding chemical, but the attempts were dismal failures.

Ref.: U. S. Patent 2,642,335; 21 CFR 121.1032, permits up to 13 ppm in table salt.

STANDARDIZED FOODS AND
FOOD LABELING

"No matter how thick
or thin you slice it
it is still boloney."

Carl Sandburg
The People, Yes[1]

One of the more puzzling and frustrating experiences for
shoppers is to encounter foods whose labels list none or
only a few of the ingredients and additives that the food
contains. Some of the foods whose labels are remarkably
blank are white and whole wheat bread, ice cream and
sherbet, jam and jelly, salad dressing and mayonnaise, and
soft drinks.

The existence of "silent labels" is attributable to the
"Definitions" and "Standards of Identity" that the Food
and Drug Administration and the Department of Agriculture have established for common foods. These official
regulations specify the ingredients and additives that must
or may be used in manufactured foods. Standards dictate such things as how much butterfat must be present
in ice cream, which (and how much) emulsifiers may be
used in bread, how much vegetable oil must be present
in salad dressing, how much caffeine and quinine may
be used in soft drinks. The regulations also indicate
whether food additives must be identified on the label
by a specific name ("mono- and diglycerides"), by a
generic name ("emulsifier added"), or not at all. Manu-

[1] Courtesy of Harcourt Brace Jovanovich, Inc.

facturers must list all the ingredients in foods for which the government has not set standards.[2]

The rationale behind food standards is a good one: a shopper should be able to assume that foods meet certain minimum standards of composition. In the absence of standards an unscrupulous manufacturer could legally call ice milk "ice cream" or could label a watered-down orange drink "orange juice" or could leave the vitamins out of its "enriched" bread. Standards outlaw these kinds of malpractices by defining what a food is.

There are, however, at least three general difficulties with food standards. First, a standard may be established so as to suit the industry concerned, with the chips falling where they may for the general public. Standards are ordinarily based on prevailing industry practices. For example, the Department of Agriculture set the meat content of corned beef hash at 35 percent—not 30, 40, 50, or 60 percent—because that was the average amount used by meatpackers. Moreover, the standard did not require a minimum protein content, because that might have interfered with the use of fatty meat. Thus, the criteria for setting the standard were commercial, not nutritional. It's anybody's guess what the composition of this product would have been if informed and aggressive consumers had been involved in the decision-making process; perhaps corned beef hash would be twice as nutritious as the corned beef trash now on supermarket shelves.

In 1969 the public was shocked into action by the revelation that the hot dog had gradually deteriorated from a food containing 20 percent protein and 19 percent fat

[2] The law requires that ingredients be listed on the label in order of decreasing predominance, by weight. Thus, a sausage that is made of "beef, water, pork, and sodium nitrite" contains more beef than water, more water than pork, and more pork than sodium nitrite.

to an item containing 12 percent protein and 35 percent fat. The reasons for the huge increase in fat content were succinctly described in Department of Agriculture hearings by the U. S. Department of Agriculture's Dr. Jack C. Leighty, Director of Technical Services Division, Consumer and Marketing Service. Dr. Leighty said:

> The incentive to incorporate more fat has been twofold. Very lean franks are somewhat tough when cooked. Increasing the amount of fat to a reasonable level makes the product more tender, improving palatability. *However, a major incentive has probably been the reduction in raw material costs achieved by using cuts of meat containing more fat and less protein* [italics added].

In the hearings the public and the hot dog manufacturers had a battle royal over the fat content, with the public demanding a limit of 30 percent or less and manufacturers pleading for no limit at all, or, at the very least, 33 percent. This was the first time that public hearings had been held, and, for the first time, a dialogue was established and the public's interests triumphed. Whether open hearings and public participation (and triumphs) remain the exception or become the rule will determine whether food standards work for or against the public interest.[3]

The second problem with Standards of Identity for many foods is that the law compels manufacturers to de-

---

[3] The hot dog victory was, unfortunately, not as total as most people believe, because the standard for frankfurters—and every other meat-containing food—suffers from a serious flaw. Although some standards set minima on the meat content and maxima on the fat content, not a single standard specifies a minimum percentage of protein. The lack of protein minima is inexcusable, because the nutritive value of meat is due almost entirely to the protein content. Thus, it is perfectly legal for manufacturers to fill the meat quota of frankfurters, beef stew, and other products with meat that consists primarily of fat, gristle, and water.

clare on the labels only the name of the product and not the ingredients and additives. Manufacturers have argued that the consumer is sufficiently protected by Standards of Identity and that lists of ingredients on labels are unnecessary and even confusing. Hence, mayonnaise labels need not indicate that mayonnaise is made of eggs, vinegar, and vegetable oil and may be spiced with MSG and artificial flavoring; ice cream labels need not indicate which artificial colorings, flavorings, stabilizers, and emulsifiers are added.

The FDA booklet "Food Additives—facts for consumers" informs the public that "only safe chemical additives have been permitted in standardized foods." Generally speaking, the FDA is correct, but when you get down to individual additives and individual people FDA's logic collapses. What is safe for one person may be hazardous to another. Persons who are on low-salt diets or are allergic to a particular emulsifier or are diabetic have difficulty avoiding their nemeses when all ingredients and additives are not listed on the label.

Many consumers and an increasing number of government officials are asking that the ingredients of standardized foods be listed on the label:

—Mrs. Mary Gullberg, home economist with Consumers Cooperative of Berkeley, in testimony before the Senate Consumer Subcommittee (January 16, 1970) reported that:

> Consumers have been telling us for a long, long time . . . that they want all the ingredients listed on all the food products. It is very confusing to have standardized foods which don't have to list ingredients alongside foods that do. Not only for reasons of health, or religious reasons do they need to have this information, but it does help them to judge the quality.

—Three panels of the 1969 White House Conference on Food and Nutrition recommended that all ingredients be listed on all foods.

—The Federation of Homemakers, Arlington, Virginia, "a nationwide group of public-spirited individuals [about 5,000 dues-paying members], deems it most urgent and therefore requests that all food ingredients, including food additives, no matter how insignificant, be listed clearly in specific terms on said food labels, even when a Standard of Identity has been established, to inform the public and to safeguard its health."

—Charles C. Edwards, Commissioner, Food and Drug Administration, testified before the Senate Consumer Subcommittee (March 23, 1970) that he could see no reasons for exempting any foods from the requirement of ingredient labeling on the packages.

Defenders of incomplete labeling contend that lists of additives on labels only confuse the consumer. Actually, as Mrs. Gullberg stated, it is the current situation that is confusing. How is the consumer to know whether the absence of a list of additives on a label signifies that (a) no additives are present, or (b) the food is covered by a standard and is brimming over with additives? For example, does the soft drink not contain caffeine or does a standard permit caffeine in the drink but not require disclosure on the label? Does the yogurt not contain artificial thickening agents, or is there a standard that permits or even requires such additives to be used without being listed on the label? The way to eliminate such confusion is obvious: all food labels should bear a complete list of ingredients.[4]

[4] The decision to purchase a food is frequently made on the basis of information supplied in an advertisement. For this reason, advertisements in all media for all manufactured or processed foods should include complete lists of ingredients.

In 1971 a group of George Washington University Law School students, organized under the acronym LABEL, exerted an enormous effort to persuade the FDA to require the complete labeling of standardized foods. They brought the matter to the public's attention via the media, talked repeatedly with government officials, and finally filed a formal petition with the FDA. In the end, though, the FDA rejected their petition on the grounds that the law did not give them the authority to require the complete labeling of those foods.

Giant Foods, a supermarket chain in the Washington area, has taken the lead in listing the ingredients on its private label standardized products. A few manufacturers, such as Coca-Cola, also see the writing on the wall and are listing ingredients voluntarily.

Whether a new law is passed by Congress or the present law is interpreted more liberally, it appears inevitable that in the near future standardized foods will bear complete lists of ingredients. As nice as these lists will be, however, we should not lose sight of the fact that they will not make white bread any more nutritious, soft drinks any less harmful, and ice cream any purer; an overpriced, adulterated, artificially colored and preserved food is not mended by a label.

The third problem with food standards is that they are so easy to get around. If a manufacturer wants to make something that is similar to a standardized food, but that contains filler, nutrients, or other ingredients not permitted by the standard, all he has to do is call the product by another name. One food standard, for instance, assures consumers that a jar labeled "peanut butter" contains at least 90 percent peanuts. A manufacturer who wishes to market peanut butter composed of 50 percent peanuts and 50 percent filler could do so simply by calling his product "imitation peanut butter" or "peanut swirl."

To let the shopper know that it is to be used like peanut butter, it would be artificially colored to the same hue as real peanut butter and packaged in a peanut butter jar, perhaps with a label declaring "great with jelly!" In other words, if your product doesn't meet the standard, just call it something else.

Standards of Identity established by the FDA are published in the Code of Federal Regulations (CFR), Title 21, parts 10–85. Those set by the Department of Agriculture are in CFR Title 7, part 81 (poultry), and Title 9, part 319 (meat). These volumes will be found in law libraries and many public libraries or may be purchased from the Government Printing Office in Washington, D.C., for $1.75, $3, and $2, respectively. But before you run out and order the Code of Federal Regulations, and despite the fact that the FDA assures us that "consumers can depend upon basic composition" of standardized foods, you should be forewarned that even with the official U. S. Government food standards at your side, you *still* could not determine the exact composition of your dish of ice cream or loaf of white bread. The reason for this strange circumstance is that Standards of Identity list not only mandatory ingredients, but also optional ingredients and additives. In other words, the standards reveal what *may* be, but are often *not necessarily,* added to foods. For instance, according to the regulation, ice cream may contain up to 0.5 percent propylene glycol alginate, but the reader's brand may have none; instead the reader's ice cream may contain 0.5 percent sodium carboxymethylcellulose or, incredible as it may be, *no* artificial stabilizer at all! Carbonated beverages may—or may not—contain brominated vegetable oil. In many other cases, additives are simply lumped under the phrases "artificial coloring" and "artificial flavoring."

Despite their ambiguities, food standards do make interesting reading and give one a pretty fair picture of what many common foods contain. The following pages contain condensations of the official definitions of a variety of food products. For the reader's convenience and sanity, the regulations have been translated from legalese into English. When a dotted line is present, items above the line are mandatory, and usually major, ingredients; items below the line are optional ingredients. Except where noted, ingredients and additives do not have to be listed on the label. When the function of an additive is not described, refer to the individual listings in Chapter 2 or Appendix 2. Unfamiliar words, phrases, and abbreviations are explained in the Glossary (Appendix 4).

FOOD STANDARDS OF IDENTITY

Chicken (or Turkey) Products[5]

	minimum percentage chicken in final product (cooked, with skin, fat, seasoning)	maximum percentage moisture that may be added (water, broth)
1. boned—solid pack	95	5
2. boned	90	10
3. boned with broth	80	20
4. boned with ____% broth	50	50
5. strained or chopped, with broth	43	57
6. "high meat" dinner (baby food)	18.75	

	minimum percentage chicken (cooked, deboned)
chicken ravioli	2
chicken soup (ready to serve basis)	2
chop suey with chicken	2

[5] 7 CFR 81.157-8, 81.167.

chicken chop suey 4
chicken chow mein without noodles 4
chicken tamales 6
noodles or dumplings with chicken 6
chicken stew 12
chicken pie 14
frozen chicken dinner 18
creamed chicken 20
chicken cacciatore 20
chicken fricassee 20
chicken à la king 20
sliced chicken with gravy and dressing 25
sliced chicken with gravy 35
minced chicken barbecue 40

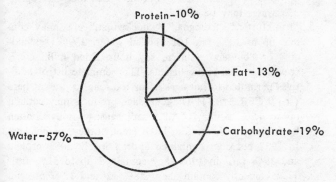

CHICKEN POT PIE

Meat Products[6]

1. "chopped beef," "ground beef," and "hamburger" consist of chopped fresh or frozen beef with or without

[6] 9 CFR 319. Meat consists primarily of water (45 to 70%, the fattier the meat, the lower the moisture content), protein (up to 20%), and fat. Thus, when a certain percentage of "meat" is specified in a food standard, the *protein* content of the food is actually a much lower percentage.

the addition of beef fat and/or seasoning, and must not contain added water, binders, or filler. These meats may contain up to 30% fat. Consumers' Union has recommended that this limit be lowered to 25% (*Consumer Reports,* August 1971).

2. "beef patties": there is no limit on the fat content of the beef; water may be added.

3. "breakfast sausage" is made from meat and meat by-products that contain up to 50% easily trimmed fat (there is no limit on the fat that grains the meat and would not be easily trimmed). Up to 3% water and 3.5% "binder" or "extenders" (starch, soy protein, nonfat dry milk, etc.) may be added.

4. "smoked pork sausage": made from pork that contains up to 50% "trimmable," but not removed, fat; up to 3% water may be added.

5. "frankfurter," "bologna," "knockwurst," etc., may contain up to 30% fat, 10% added water, 3½% "extenders" or "binders" (starch, soy flour, dried milk, etc.), seasonings, and curing agents. The complete list of additives permitted in sausages is far too long to present here (see 9 CFR 318.7). These sausage products may contain up to 15% poultry. "All meat" sausages may contain beef, pork, veal, mutton, lamb, goat, turkey, or chicken. In 1937 hot dogs contained approximately 20% protein and 19% fat;[7] in 1969 they contained up to 51% fat![8] They currently contain about 28% fat and 12% protein.

Every friend or foe of frankfurters ought to take a tour through a local packinghouse, if they will let you in. My visit to a large, Virginia packinghouse was a real eye-opener. The meat that went into hot dogs (and bologna) consisted entirely of fat trimmed from hams and chops; the only red meat was bits that were accidentally attached to large pieces of fat.

[7] Testimony of Dr. R. K. Somers, U. S. Dept. of Agr., before the House Government Operations Committee, June 4, 1969. See also *Consumer Reports,* February 1972, page 73.

[8] Testimony of Dr. Jack C. Leighty, U. S. Dept. of Agr., at USDA hearings, June 18, 1969.

The low protein content of such meat is often boosted slightly by the addition of dry milk or soy flour. After the ingredients were pulverized and emulsified, the mixture of fat, water, protein and additives had the color of bread dough and the consistency of mud. The emulsion was then packed into a casing, which was coated with a brilliant red artificial color-

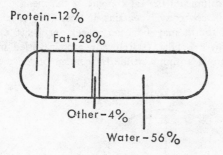

HOT DOG

ing, and cooked. The light tan mixture darkened during cooking due to the heat and to the action of sodium nitrate and nitrite. In addition, the red dye on the casing colored the surface of the frankfurter. Additives like sodium erythorbate, sodium ascorbate, and sodium acid pyrophosphate sped up the development of color. A visit to their local frankfurter factory will cure most persons of their hot dog habit.

Most brands of hot dogs are made from muscle meat and fat, but some manufacturers throw in everything but the pig's squeal and the cow's moo. One brand that is sold two miles from the U. S. Capitol is constructed from beef, pork stomachs, unskinned pork jowls, pork salivary glands, lymph nodes and fat (cheek), pork spleens, isolated soy protein, sodium erythorbate, sodium nitrate, sodium nitrite, and artificial coloring. (Federal regulations stipulate that meat by-products and chemical additives be listed on the label.) Salivary glands and sodium nitrite notwithstanding, Lawrence Russell, the marketing vice-president of Tee-Pak, Inc., recently had the audacity to criticize frankfurter producers for not educating consumers to the fact that franks are pure, economical, and convenient sources of nutrition! Mr. Russell called for "educational" campaigns and the development of new products, such as "pizza franks" and "ski franks."[9]

6. "luncheon meat" and "meat loaf" are made from finely chopped meat to which 3% water may be added. There is no limit on the fat content of the meat.

7. "chili con carne" contains at least 40% meat, of which one fourth may be head, cheek, or heart meat,[10] and up to 8% filler (starch, soy protein, dried milk, etc.). There is no limit on the fat content of the meat.

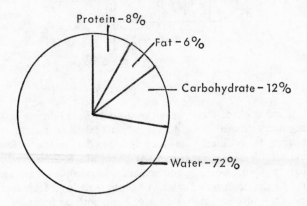

Protein – 8%
Fat – 6%
Carbohydrate – 12%
Water – 72%

CHILI CON CARNE

8. "chili con carne with beans" contains at least 25% meat, one fourth of which may be head, cheek, or heart meat.[10] There is no limit on the fat content of the meat.

9. "corned beef hash" contains at least 35% beef (5% of which may be cheek, heart, or head meat[10]), potatoes, seasonings, and curing agents. The finished product may contain up to 15% fat and 72% moisture, which does not leave much room for protein and carbohydrate.

[9] *Food Product Development,* page 96, June/July 1971.
[10] Must be noted on label, if used in the product.

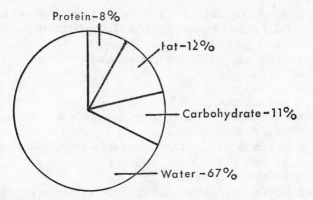

CORNED BEEF HASH

10. "beef stew" contains not less than 25% meat, but there is no limit on the fat content of the meat.

11. "spaghetti with meatballs in sauce" contains at least 12% meat, but there is no limit on the fat content of the meat. The meatballs may contain up to 12% binder (soy flour, dried milk, starch, etc.).

12. "spaghetti sauce with meat" contains at least 6% meat, but there is no limit on the fat content of the meat.

13. "beef pie," "pork pie," and similar products must contain at least 25% meat, but there is no limit on the fat content of the meat.

14. "pizza with meat" is a bread-based food product made with tomato sauce, cheese, and at least 15% meat topping. There is no limit on the fat content of the meat.

15. "deviled ham" is made from finely chopped ham and condiments; it may contain up to 35% fat.

Margarine[11]

Ingredients:

1. animal or vegetable fat or oil which may be hydrogenated

[11] All ingredients, including the source of fat or oil, must be noted on the label. Preservatives, artificial coloring, and artificial flavoring must be identified as such.

(if only vegetable oil is used, the margarine may contain finely ground soybeans); the finished product must contain at least 80% fat.

2. cream, milk, nonfat dry milk, skim milk, or water.

. . . .

3. artificial coloring.
4. a preservative: sodium benzoate or potassium sorbate, up to 0.1%.
5. vitamin A (at least 15,000 units per pound), natural or synthetic, with or without vitamin D.
6. artificial butter flavoring.
7. lecithin and/or mono- and diglycerides (may be the sulfoacetate derivative), up to 0.5%.
8. butter.
9. citric acid, isopropyl citrate (up to 0.02%), stearyl citrate (0.15%), calcium disodium EDTA (up to 75 ppm), or BHA and/or BHT (up to 0.02%) to serve as an antioxidant.
10. salt.

Milk Chocolate[12]

Ingredients:

1. chocolate liquor, made by grinding shelled cacao beans. The finished product contains at least 10% (by weight) of this ingredient.
2. a dairy product: milk, concentrated milk, skim milk, nonfat dry milk, etc. The finished product contains at least 3.66% milk fat and at least 12% milk solids.
3. a sweetener: sugar, dried corn syrup, dextrose, etc.

. . . .

4. spice or flavoring: ground spice, ground vanilla beans, honey, salt, dried malted cereal extract, any natural or artificial[13] (including vanillin and ethyl vanillin) flavor-

[12] 21 CFR 14.7.
[13] Must be noted on label.

ing; flavorings that imitate the taste of chocolate, milk, or butter may not be used.

5. emulsifier(s): lecithin, with or without related phosphatides, (0.5%) to reduce the viscosity of chocolate and to replace some of the cocoa butter; mono- and diglycerides, or a phosphate derivative thereof (up to 0.5%); sorbitan monostearate (up to 1%); polysorbate 60 (up to 0.5%). Use of an emulsifier must be noted on the product label.

Flour[14]

Ingredients:

1. wheat, with up to 0.75% malted barley flour.

. . . .

2. ascorbic acid as a dough conditioner (up to 200 ppm).[15]
3. one or more bleaching[16] or aging ingredients: oxides of nitrogen; chlorine; nitrosyl chloride; chlorine dioxide; one part of benzoyl peroxide mixed with not more than six parts of one or more of potassium alum, calcium sulfate, magnesium carbonate, sodium aluminum sulfate, dicalcium or tricalcium phosphate, starch, calcium carbonate; acetone peroxides; azodicarbonamide (not more than 45 ppm).

Enriched flour (21 CFR 15.10) contains specified amounts of vitamin B-1 (thiamine), B-2 (riboflavin), niacin or niacinamide, and iron; it may also contain specified amounts of vitamin D, calcium, and wheat germ. Despite all these additions, it still lacks many of the vitamins and minerals that are naturally present in whole wheat flour.

White Bread[17]

Ingredients:

1. flour (may be bromated or phosphated).
2. water.

[14] 21 CFR 15.1.
[15] Label must bear the statement "ascorbic acid added as a dough conditioner."
[16] If a bleach is used, the label must bear the word "bleached."
[17] 21 CFR 17.

3. salt.

4. shortening, to which may be added such emulsifiers as lecithin, mono- and diglycerides, diacetyl tartaric acid esters of mono- and diglycerides, propylene glycol mono- and diesters of fatty acids.

5. milk or milk product: milk, skim milk, nonfat dry milk (may contain carrageenan), buttermilk, cheese whey, milk proteins, etc.

6. egg: liquid, frozen, or dried egg, egg yolk, or egg white.

7. sugar, molasses, honey, corn syrup, invert sugar, etc.

8. an agent which can change starch into sugar and dextrin: diastatically active malt syrup or malted wheat flour, appropriate enzymes from microorganisms or pineapple.

9. dried yeast (not more than 2 parts per 100 parts of flour).

10. harmless lactic acid-producing bacteria.

11. corn, potato, rice flour; wheat, corn, or potato starch, etc. (any of which may be partly dextrinized); not more than 3 parts per 100 parts flour.

12. ground dehulled soybeans which possess enzymatic activity (up to 0.5 parts per 100 parts flour).

13. yeast food: calcium salts of lactate, sulfate, carbonate; dicalcium phosphate; ammonium salts of phosphate, sulfate, chloride (up to 0.25 parts per 100 parts flour).

14. potassium or calcium bromate, potassium or calcium iodate, calcium peroxide, azodicarbonamide (with tricalcium phosphate added to prevent caking) to serve as bleaching and aging agents.

15. vinegar, calcium or sodium propionate, sodium diacetate or lactic acid[18] may be used as mold inhibitors (up to 0.32 parts for every 100 parts of flour).

16. L-cysteine used in conjunction with aging agents (up to 0.009 parts per 100 parts flour).

17. spice.

18. calcium or sodium stearoyl-2-lactylate, lactylic stearate, sodium stearyl fumarate, succinylated monoglycerides, ethoxylated mono- and diglycerides, polysorbate 60

[18] Use of sodium diacetate or lactic acid must be noted on label.

(total up to 0.5 parts per 100 parts flour). These are conditioners which make drier, easier-to-machine dough and breads which have a more palatable crumb (the part of the bread that is not crust is crumb). Several of these emulsifiers inhibit bread from getting stale (which is a result of starch crystallization), thereby increasing the shelf life and saving bakers money.

19. wheat gluten, which must contain at least 75% protein (not more than 2 parts per 100 parts flour).

Enriched White Bread

Same as white bread except that it contains specified amounts of thiamine (vitamin B-1), riboflavin (vitamin B-2), niacin or niacinamide, and iron, and may contain vitamin D, calcium, wheat germ, extra milk. Despite the addition of these nutrients, enriched white bread is still deficient in fiber, magnesium, vitamin B-6, and other vitamins and minerals.

Milk Bread

Same as white bread except that it contains at least 8.2 parts by weight of milk solids for each 100 parts by weight of flour.

Raisin Bread

Similar to white bread except that it contains at least 50 parts by weight of seeded or seedless raisins for each 100 parts by weight of flour, and may bear icing or frosting.

Whole Wheat Bread

Similar to white bread except that the dough is made from whole wheat flour, which—on balance— is more nutritious than enriched white flour.

Ice Cream[19]

Ingredients:

1. dairy ingredients: cream, milk, skim milk, nonfat dry

[19] 21 CFR 20.1; see also *Federal Register 25* 7126, July 27, 1960; *Washington Post* (*Potomac* Magazine), August 22, 1971.

milk, butter, etc. The final product must contain at least 8% (by weight) milk fat and at least 16% (by weight) total milk solids, and must weigh no less than 4.5 pounds per gallon. Ice cream that weighs 4.5 pounds per gallon contains approximately 50% air! Premium ice cream weighs upward of 5.5 pounds per gallon. The greater the fat content, the smoother and richer the ice cream.

2. sweetening ingredient: sugar, dextrose, honey, corn syrup, etc.
3. flavorings: salt, spice, ground vanilla beans, any natural food flavoring; chocolate or cocoa as powder or syrup, and which may contain disodium phosphate or sodium citrate.

. . . .

4. artificial food flavoring (must be noted on label). The product is called:
   (a) "vanilla [or other flavor] ice cream" if no artificial flavors are added;
   (b) "vanilla flavored ice cream" if it contains both natural and artificial flavors but the natural flavor predominates;
   (c) "artificially flavored vanilla ice cream" or "artificial vanilla ice cream" if it contains both natural and artificial flavors and the artificial flavor predominates, or if it contains only artificial flavors.
5. natural or artificial coloring (need not be noted on label).[20]
6. fruit or fruit juice, which may be sweetened, thickened with pectin or some other agent, and acidified with citric, ascorbic, or phosphoric acid.
7. nut meats, which may be roasted, cooked in oil, preserved in syrup, and salted.
8. malted milk.

[20] Artificial coloring may be added to cheese, butter, and ice cream without being listed on the label. This provision of the Food, Drug and Cosmetic Act (section 403k) reflects the strength of the dairy industry's lobby.

9. confectionery (candy, cakes, cookies, glacéed fruit).
10. cooked cereal.
11. distilled alcoholic beverage, including wines or liqueurs, in an amount not to exceed that required to flavor the ice cream.
12. casein (the major protein in milk).
13. liquid, frozen, or dried egg or egg yolk (up to 1.4%). The use of egg facilitates the incorporation of air in commercially made ice cream.
14. thickening and stabilizing agents: agar, carrageenan, gelatin, gum tragacanth, lecithin, carboxymethylcellulose (cellulose gum), calcium sulfate, propylene glycol alginate, etc. (up to 0.5%); microcrystalline cellulose (up to 1.5%; use must be noted on label). These chemicals discourage the formation of ice crystals, produce a smoothness suggestive of richness (richness is normally a reflection of the fat content), and retard melting. Dioctyl sodium sulfosuccinate may be used to help these chemicals dissolve.
15. emulsifiers: polysorbate 65 or 80 (up to 0.1%), mono- and diglycerides (up to 0.2%). These chemicals decrease the size of fat particles; this results in a smoother, "drier" product, imitating the effect of a high cream content.
16. sodium citrate, disodium phosphate, tetrasodium pyrophosphate, sodium hexametaphosphate (up to 0.2%); calcium or magnesium carbonate, calcium or magnesium hydroxide, calcium or magnesium oxide (up to 0.04%). These additives are used ostensibly to maintain the uniformity of ice cream from batch to batch; their presence may reflect the use of slightly soured milk or cream.

Fruit Sherbet[21]

Ingredients:

1. fruit or fruit juice (may be acidified with citric, ascorbic, or phosphoric acid and thickened with pectin): not less than 2 to 10% depending upon the fruit.

[21] 21 CFR 20.4; see *Federal Register 25* 7126, July 27, 1960.

2. a dairy product: cream, milk, skim milk, nonfat dry milk, etc. The finished product must contain 1 to 2% milk fat and 2 to 5% milk solids, and must weigh no less than 3 pounds per half gallon.

3. a sweetener: sugar, corn syrup, dextrose, invert sugar, etc.

. . . .

4. egg, egg yolk (liquid, dried, or frozen), not more than 0.5% by weight of final product.

5. agar, carrageenan, guar gum, propylene glycol alginate, carboxymethylcellulose, etc. (up to 0.5% by weight of the sherbet); microcrystalline cellulose (1.5% of sherbet). These thickening agents discourage the formation of ice crystals and produce a smoothness suggestive of richness (richness is normally a reflection of fat content). Dioctyl sodium sulfosuccinate may be used to help these chemicals dissolve.

6. mono- or diglycerides (up to 0.2%); polysorbate 65 or 80 (up to 0.1%). These emulsifiers decrease the size of fat globules and produce a smoother product.

7. casein, the major milk protein.

8. an acid, to increase tartness: citric, tartaric, malic, lactic, ascorbic, phosphoric.

9. natural food flavoring.

10. artificial food flavoring (use must be noted on label).

11. coloring, including artificial (use must be noted on label).

12. salt.

Mayonnaise[22]

Ingredients:

1. vegetable oil, not less than 65% by weight of mayonnaise (the oil may contain up to 0.125% oxystearin to prevent the oil from clouding up at refrigerator temperatures).

2. vinegar (may contain up to 25% citric acid but must be

[22] 21 CFR 25.1.

noted on label) or lemon or lime juice, not less than
2.5% by weight of mayonnaise.

3. egg or egg yolk (frozen, liquid, or dried).

• • • • • • • • • • • • • •

4. salt.
5. a sweetener: sugar, honey, corn syrup, dextrose, etc.
6. any spices or natural, harmless food flavorings, provided
   they do not impart a yellow color to the mayonnaise
   (and suggesting that extra egg is present).
7. monosodium glutamate (MSG) as a flavor enhancer.
8. calcium disodium EDTA or disodium EDTA to absorb
   metal atoms, which would otherwise help the vegetable
   oil ingredient turn rancid (up to 75 ppm EDTA may be
   used). Use of EDTA must be noted on package label.

Salad Dressing[23]

Ingredients:

1. vegetable oil, at least 30% by weight of final product
   (up to 0.125% oxystearin may be added to the oil to
   prevent it from clouding up at refrigerator temperatures).
2. vinegar (may contain up to 25% citric acid and must
   be noted on the label) or lemon or lime juice.
3. liquid, frozen, or dried egg or egg yolk (egg yolk equivalent must comprise at least 4% of the weight of the salad
   dressing). Egg thickens the product and prevents the oil
   and vinegar from separating out into two layers.
4. cooked or partly cooked paste made from water plus a
   food starch, tapioca flour, wheat flour, or rye flour.

• • • • • • • • • • • • • •

5. salt.
6. sugar, corn syrup, honey, invert sugar, etc.
7. spices, except those which would impart a color simulating egg yolk.
8. monosodium glutamate (MSG) to bring out the flavor
   of the food the salad dressing is used on.

[23] 21 CFR 25.3.

9. suitable, harmless food seasoning or flavoring, except those that would impart a color simulating egg yolk.
10. emulsifying or thickening agents (total not more than 0.75% by weight of completed salad dressing and must be listed on label): gum acacia, guar gum, gum tragacanth, pectin, sodium carboxymethylcellulose, etc. Dioctyl sodium sulfosuccinate may be added to any one of these chemicals to help them dissolve (up to 0.5% of emulsifier).
11. calcium disodium EDTA or disodium EDTA (up to 75 ppm) may be added to absorb metal atoms which would otherwise help the vegetable oil turn rancid. Use must be noted on package label.

Preserves, Jam[24]

Ingredients:

1. a fruit ingredient, at least 45 parts fruit for each 55 parts sweetener. The name of the fruit or fruits must be included in the name of the product.
2. a sweetener: sugar, invert sugar syrup, dextrose, corn syrup, honey (the presence of honey must be noted on product label).

· ·

3. spice (use must be noted on label).
4. an acidifying agent (to compensate for insufficiently acidic fruit): vinegar, lemon or lime juice, citric acid, fumaric acid.
5. pectin, if needed, to compensate for a deficiency of pectin in the fruit; pectin acts as a thickening agent.
6. sodium citrate, sodium potassium tartrate (up to 3 oz. for each 100 pounds of sweetener).
7. sodium benzoate or benzoic acid as a preservative (must be labeled).
8. an antifoaming agent: butter, margarine, lard, cottonseed oil, mono- and diglycerides, etc.

[24] 21 CFR 29.3.

Soda Water[25]

Ingredients:

1. carbon dioxide dissolved in water.
2. nutritive sweetener: sugar, dextrose, corn syrup, fructose, etc.
3. fruit juice or artificial flavoring (this ingredient may be dissolved in ethyl alcohol, glycerin, or propylene glycol). The name of the fruit must be incorporated into the product's name. The presence of artificial flavoring must be noted on the label.
4. natural and artificial colorings (the latter must be noted on the label).
5. an acid to add tartness: acetic, adipic, citric, fumaric, gluconic, lactic, malic, phosphoric or tartaric acid.
6. a buffering agent to maintain the desired acidity: one or more of the acetate, bicarbonate, carbonate, chloride, citrate, gluconate, lactate, phosphate or sulfate salts of calcium, magnesium, potassium, or sodium.
7. one or more emulsifying, stabilizing, or thickening agents: brominated vegetable oil, guar gum, lecithin, pectin, sodium alginate, carboxymethylcellulose, hydroxylated lecithin, sodium hexametaphosphate, etc. These ingredients may contain up to 0.5% dioctyl sodium sulfosuccinate to help them dissolve. These substances add "body" to soda and/or keep flavor oils from separating out.
8. a foaming agent: ammoniated glycyrrhizin, gum ghatti, licorice, yucca, or quillaia (used in root beer).
9. quinine (not to exceed 83 ppm of finished beverage),

---

[25] 21 CFR 31.1. This regulation does not cover diet drinks, so one can discover the ingredients of many soft drinks by holding a magnifying glass to the bottle tops of diet drinks. "Fresca" contains carbonated water, citric acid (flavor and tartness), saccharin (sweetener), sodium citrate (buffer), gum arabic (adds "body"), natural and artificial flavoring, brominated vegetable oil (emulsifier), salt, artificial coloring, sodium benzoate (preservative), and stannous chloride (antioxidant); "Diet Pepsi" contains carbonated water, sugar, caramel coloring, phosphoric acid, saccharin, caffeine, flavorings, and citric acid.

      an extremely bitter substance. Use must be noted on label.

10. an antioxidant to prevent off-colors or off-flavors from developing: ascorbic acid, BHA, BHT, propyl gallate, sodium bisulfite, tocopherols; stannous chloride (up to 11 ppm) in canned beverages.

11. a preservative to prevent the growth of microorganisms: methyl or propyl paraben, sodium benzoate, sorbic acid.

12. the defoaming agent dimethyl polysiloxane (up to 10 ppm).

13. caffeine (up to 0.02%, approximately half the amount in coffee), which acts as a stimulant. The presence of caffeine must be noted on the product label *except* in cola or "pepper" (Dr. Pepper) drinks, which always contain caffeine.

Breaded Shrimp[26]

Frozen Raw Breaded Shrimp must consist of at least 50% shrimp (that is, up to 50% of the product may be bread crumbs and batter).

Frozen Raw Lightly Breaded Shrimp must consist of at least 65% shrimp (that is, up to 35% of the product may be bread crumbs and batter).

The composition of the breading need not be listed on packages of frozen shrimp. Preservatives may be added to prevent discoloration of the shrimp or spoiling of oils, but they must be listed on the label. These regulations prohibit the use of artificial coloring, artificial flavoring, and artificial sweeteners.

Peanut Butter[27]

Ingredients:

1. blanched peanuts, in which the germ may or may not be included, or unblanched peanuts, including the skins and germ.

. . . .

[26] 21 CFR 36.30-31.
[27] 21 CFR 46.1.

2. up to 10% seasoning and stabilizing ingredients, such as salt and hydrogenated vegetable oil. Use of unblanched peanuts, seasoning, and stabilizing ingredients must be labeled.

Appendices

APPENDIX 1: ADDITIVES THAT HAVE BEEN BANNED

Additive	Function	Source	Last used	Reason for ban
agene (nitrogen trichloride)	flour bleaching and aging agent	synthetic	1949	Dogs that ate bread made from treated flour suffered epileptic-like fits; the toxic agent was methionine sulfoxime.
coal tar dyes	artificial coloring	synthetic		
butter yellow			1919	toxic, later found to cause liver cancer
FD&C Green 1			1966	liver cancer
FD&C Orange 1			1960	organ damage
FD&C Orange 2			1960	organ damage
FD&C Red 1			1960	liver cancer
FD&C Red 4				High levels damaged adrenal cortex of dog; after 1965 used only in maraschino cherries and certain pills.
FD&C Red 32				Damages internal organs and may be a weak carcinogen; since 1956 used under the name Citrus Red 2 to color oranges (2 ppm).
Sudan 1			1960	toxic, later found to be carcinogenic.
FD&C Yellow 1 and 2			1960	intestinal lesions at high dosages
FD&C Yellow 3			1960	heart damage at high dosages
FD&C Yellow 4			1919	heart damage at high dosages

substance	use	source	year	effect
cobalt salts	stabilize beer foam	synthetic	1966	toxic effects on heart
coumarin	flavoring	tonka bean	1954	liver poison
cyclamate	artificial sweetener	synthetic	1970	bladder cancer
diethyl pyrocarbonate (DEPC)	preservative (beverages)	synthetic	1972	combines with ammonia to form urethan, a carcinogen
dulcin (p-ethoxyphenyl urea)	artificial sweetener	synthetic	1950	liver cancer
ethylene glycol	solvent, humectant	synthetic	1941	kidney damage
monochloroacetic acid	preservative	synthetic		highly toxic
nordihydroguaiaretic acid (NDGA)	antioxidant	desert plant	1971*	kidney damage
oil of calamus	flavoring	root of calamus	1968	intestinal cancer
polyoxyethylene-8-stearate (Myrj 45)	emulsifier (used in baked goods)	synthetic	1952	high levels caused bladder stones and tumors
safrole	flavoring (root beer)	sassafras	1960	liver cancer
thiourea	preservative	synthetic	c. 1950	liver cancer

*NDGA was banned by the FDA in 1968, but the Department of Agriculture did not ban it until 1971.

APPENDIX 2: PARTIAL LIST OF COMPOUNDS "GENERALLY RECOGNIZED AS SAFE (GRAS)"[1]

Substance	Functions
acacia (gum arabic)	thickening agent
acetic acid,	acid, antimicrobial, flavoring
calcium diacetate	emulsifier salt
sodium acetate	buffer
sodium diacetate	antimicrobial preservative
adipic acid	acid, flavoring
agar	thickening agent
alanine	amino acid
alginic acid and salts	thickening agent
ammonium	
calcium	
potassium	
sodium	
aluminum compounds	firming agent, miscellaneous
ammonium sulfate	
potassium sulfate	
sodium sulfate	
sulfate	
ammonium salts	miscellaneous
bicarbonate	alkali
carbonate	leavening agent
chloride	yeast food, salt substitute
hydroxide	alkali
phosphate, dibasic	buffer
phosphate, monobasic	leavening agent
sulfate	leavening agent
arginine	amino acid
ascorbic acid (vitamin C)	nutrient, antioxidant, acid, flavoring
ascorbyl palmitate	nutrient, antioxidant
calcium ascorbate	nutrient, antioxidant
sodium ascorbate	nutrient, antioxidant
aspartic acid	amino acid
beeswax	coating on vitamin pills, candy glaze
bentonite	filtering aid

[1] Listed in section 121.101(d) of Title 21 of the Code of Federal Regulations.

Substance	Functions
benzoic acid sodium benzoate	antimicrobial preservative
biotin	vitamin
butane	gas
butylated hydroxyanisole (BHA)	antioxidant preservative
butylated hydroxytoluene (BHT)	antioxidant preservative
caffeine	stimulant
calcium salts	nutrient, firming agent, miscellaneous
acetate	sequestrant, buffer
carbonate	alkali, white coloring excipient, yeast food
chloride	firming agent in canned fruits and vegetables
gluconate	buffer, sequestrant
hexametaphosphate	sequestrant
hydroxide	alkali
phosphate (mono-, di-, tribasic)	sequestrant, acid in baking powder (mono), anti-caking agent (tribasic)
phytate	sequestrant
pyrophosphate	sequestrant
stearate	anti-caking agent
sulfate	alkali, yeast food, dough conditioner
caprylic (octanoic) acid	flavoring
caramel	coloring
caranuba wax	candy glaze
carbon dioxide	acid, carbonated drinks, gas propellant
carboxymethylcellulose	thickening agent
carob bean gum (locust bean gum)	thickening agent
carrageenan	thickening agent
caseinate, sodium	milk protein
cellulose	bulking agent

cholic acid
  desoxycholic acid
  glycocholic acid
  ox bile extract
  taurocholic acid
	emulsifier

citric acid and salts	acid, flavoring, sequestrant, nutrient

Appendix 2 (continued)

Substance	Functions
calcium citrate	buffer
isopropyl citrate	sequestrant (antioxidant)
manganese citrate	nutrient
monoisopropyl citrate	sequestrant (antioxidant)
potassium citrate	
sodium citrate	
stearyl citrate	sequestrant (antioxidant)
citrus bioflavonoids (naringin, hesperidin)	preservative, nutrient (vitamin P)
copper gluconate	nutrient
cuprous iodide	nutrient (used in table salt)
cysteine	amino acid, in meat flavoring
cystine	amino acid
dextrans	thickening agent
diacetyl tartaric acid ester of mono- and diglycerides	emulsifier
erythorbic acid	antioxidant
ethyl formate	flavoring, antimicrobial
ghatti gum	thickening agent
glucono-delta-lactone	leavening, acid, in meat curing
glutamic acid and salts	amino acid, flavor enhancer
monoammonium	
monopotassium	
monosodium	
glycerin (glycerol)	solvent, humectant
glycerophosphate salts	nutrient
calcium	
manganese	
potassium	
glyceryl triacetate (triacetin)	wax-like coating
glycine	amino acid, flavor enhancer
guar gum	thickening agent
gum guaiac	antioxidant
helium	gas propellant
histidine	amino acid
hydrochloric acid	acid
hydrogen peroxide	bleach
hydrolyzed vegetable protein (HVP)	flavor enhancer
hydroxy benzoates	antimicrobial preservative
methyl ester	
propyl ester	

Substance	Functions
inositol	vitamin
iron compounds	nutrient
ferric ammonium citrate	
ferric phosphate	
ferric pyrophosphate	
ferric sodium pyrophosphate	
ferrous citrate	
ferrous gluconate	artificial coloring in black olives
ferrous lactate	
ferrous sulfate	
reduced iron	
isoleucine	amino acid
karaya gum	thickening agent
lactic acid	acid, flavoring
calcium lactate	nutrient, yeast food
lecithin	emulsifier, antioxidant, nutrient
monosodium phosphate derivatives of mono- and diglycerides; choline chloride	
leucine	amino acid
linoleic acid	vitamin
lysine	amino acid
magnesium salts	nutrient, miscellaneous
carbonate	anti-caking agent, alkali
chloride	
hydroxide	alkali
oxide	alkali
phosphate (di- and tribasic)	sequestrant
stearate	anti-caking agent
sulfate	water corrective (brewing industry)
malic acid	acid, flavoring
manganese salts	nutrient
chloride	
gluconate	
hypophosphite	
oxide	
sulfate	
mannitol	sweetener, humectant, dust on chewing gum
methyl cellulose	thickening agent
mono- and diglycerides	emulsifier, anti-staling agent in bread
niacin, niacinamide	vitamin

Appendix 2 (continued)

Substance	Functions
nitrogen	gas-packed foods
nitrous oxide	whipping agent and gas propellant
pantothenate	vitamin
calcium salt	
pantothenyl alcohol	
sodium salt	
papain	meat tenderizer
pectinate, sodium	thickening agent
phenylalanine	amino acid
phosphoric acid	acid, sequestrant
potassium salts	miscellaneous
bicarbonate	
bromate	flour treatment agent
carbonate	alkali
chloride	salt substitute, yeast food
hydroxide	alkali
iodate	nutrient, flour treatment agent
iodide	nutrient
nitrate	meat curing agent
nitrite	meat curing agent
phosphate (dibasic)	sequestrant
sulfate	water corrective (brewing industry)
proline	amino acid
propane	gas
propionic acid	antimicrobial preservative
calcium propionate	
sodium propionate	
propyl gallate	antioxidant preservative
propylene glycol	solvent, humectant
pyridoxine hydrochloride	vitamin
rennin	enzyme used in making cheese
riboflavin	vitamin B-2
riboflavin-5-phosphate	
serine	amino acid
silicates	anti-caking agents
aluminum calcium	
calcium	
magnesium	
sodium alumino	
sodium calcium alumino	
silica aerogel	
tricalcium	

Substance	Functions
smoke flavor solution	flavoring
sodium salts	miscellaneous
aluminum phosphate	acid in baking powder
bicarbonate	baking soda
carbonate	alkali
gluconate	sequestrant
hexametaphosphate	sequestrant
metaphosphate	sequestrant
phosphate (mono-, di- and tribasic)	sequestrant
pyrophosphate	sequestrant
pyrophosphate (tetra)	sequestrant
sesquicarbonate	
thiosulfate	antioxidant, sequestrant
tripolyphosphate	sequestrant
sorbic acid and salts	antimicrobial preservative
calcium	
potassium	
sodium	
sorbitol	humectant, sweetener
stannous chloride	antioxidant preservative
succinic acid	acid
sulfur dioxide	preservative
potassium bisulfite	
potassium metabisulfite	
sodium bisulfite	
sodium metabisulfite	
sodium sulfite	
sulfuric acid	acid
tartaric acid and salts	acid, flavoring
potassium (cream of tartar)	acid in baking powder, buffer
sodium	
sodium potassium (Rochelle salts)	buffer
thiamine hydrochloride	vitamin B-1
thiamine mononitrate	
thiodipropionic acid	antioxidant preservative
dilauryl thiodipropionate	antioxidant preservative
threonine	amino acid
tocopherols	vitamin E
tocopherol acetate	vitamin E
tragacanth gum	thickening agent
triethyl citrate	flavoring
tryptophan	amino acid
tyrosine	amino acid

Appendix 2 (continued)

Substance	Functions
valine	amino acid
vitamin A	nutrient
vitamin A acetate	nutrient
vitamin A palmitate	nutrient
carotene	nutrient, coloring
vitamin B-12	nutrient
vitamin D-2, D-3	nutrient
zinc salts	nutrient
chloride	
gluconate	
oxide	
stearate	
sulfate	

APPENDIX 3:
CHEMICAL FORMULAS OF ADDITIVES

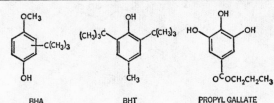

BHA

BHT

PROPYL GALLATE

SACCHARIN

CYCLAMATE

VANILLIN (R = CH₃)
ETHYL VANILLIN (R = CH₂CH₃)

CAFFEINE

MALTOL (R = CH₃)
ETHYL MALTOL (R = CH₂CH₃)

SODIUM BENZOATE

METHYL PARABEN

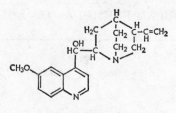

QUININE

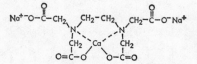

CALCIUM DISODIUM EDTA

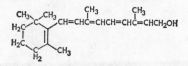

VITAMIN A

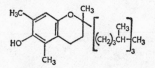

ALPHA TOCOPHEROL (VITAMIN E)

S(CH₂CH₂COOH)₂ $\quad$ $\begin{array}{c} H_2COH \\ HCOH \\ CH_3 \end{array}$ $\quad$ HOOC(CH₂)₄COOH

THIODIPROPIONIC ACID $\qquad$ PROPYLENE GLYCOL $\qquad$ ADIPIC ACID

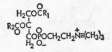

LECITHIN

SODIUM STEAROYL LACTYLATE

CITRIC ACID (R = H)
STEARYL CITRATE (R = H OR STEARYL)
ISOPROPYL CITRATE (R = H OR 2-PROPANOL)

DIMETHYLPOLYSILOXANE

TARTARIC ACID

MALIC ACID

LACTIC ACID

(l)

(d)

FUMARIC ACID

R_1, R_2, R_3 = fatty acids

R_1, R_2 = fatty acids

GLYCERIN

VEGETABLE OIL

MONOGLYCERIDE

DIGLYCERIDE

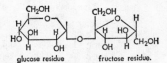

glucose residue

fructose residue.

SUCROSE (SUGAR)

$(CH_3CH_2COO^-)_2Ca^{++}$

$H_3C-CH=CH-CH=C-COOH$

CALCIUM PROPIONATE

SORBIC ACID

$H_2N\overset{O}{\underset{}{C}}N=N\overset{O}{\underset{}{C}}NH_2$

AZODICARBONAMIDE

BENZOYL PEROXIDE

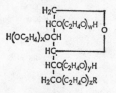

R=oleic acid
w+x+y+z = approx. 20

POLYSORBATE 80

$\begin{array}{l} O=COCH_2CH(CH_2)_3CH_3 \\ \quad\quad\quad\quad C_2H_5 \\ H_2C \\ HCSO_3^-\ Na^+ \\ \quOCH_2CH(CH_2)_3CH_3 \\ O=C \\ \quad\quad\quad\quad C_2H_5 \end{array}$

DIOCTYL SODIUM
SULFOSUCCINATE

$\begin{array}{l} CH_2OH \\ HOCH \\ HOCH \\ HCOH \\ HCOH \\ CH_2OH \end{array}$

MANNITOL

$\begin{array}{l} CH_2OH \\ HCOH \\ HOCH \\ HCOH \\ HCOH \\ CH_2OH \end{array}$

SORBITOL

$\begin{array}{l} COOH \\ HCOH \\ HOCH \\ HCOH \\ HCOH \\ CH_2OH \end{array}$

GLUCONIC ACID

ERYTHORBIC ACID

ASCORBIC ACID
(VITAMIN C)

R=H or stearic acid

SORBITAN MONOSTEARATE

APPENDIX 4: GLOSSARY

acids (acidulants): serve many food uses, as flavorings (citric acid —citrus; malic acid—apple; acetic acid—vinegar; tartaric acid— grape), preservatives (acidity prevents the growth of micro- organisms), and antioxidants (molecules with two acidic groups, such as citric acid, may be used to trap metal ions that might otherwise cause food to discolor or go rancid). Acids are used to adjust and stabilize the acidity of cheese and beer to optimize growing conditions for microorganisms. Carbonic acid is used to make carbonated water. See Chapter 2 for individual acids.

aging (maturing) agent: a chemical added to flour or dough that oxidizes the protein. Freshly milled, untreated flour makes sticky, poor-rising dough and low-quality bread and rolls. In the old days, millers aged flour by holding it in bins for several months and letting the oxygen in the air act on it. The aging causes changes in the protein in flour that make the dough elastic enough to let gas bubbles (generated by yeast) form and rigid enough to hold the bubbles in the dough. Bakers now use chemicals to age flour. Chemical aging requires much less time and space, offers little opportunity for vermin to consume or contaminate the flour, and results in batches with more consistent properties as compared to natural aging. See acetone peroxide, azodicarbonamide, chlorine, chlorine dioxide, potassium bromate, potassium iodate. Some of these agents also bleach flour.

amino acids: the sub-units from which proteins are built. There are twenty major amino acids: alanine, arginine, asparagine, aspartic acid, cysteine, glutamic acid, glutamine, glycine (amino- acetic acid), histidine, isoleucine, leucine, lysine, methionine, phenylalanine, proline, serine, threonine, tryptophan, tyrosine, valine.

anticaking compounds: chemicals that manufacturers add to powdered or granular foods to prevent them from absorbing moisture and becoming lumpy. See silicates, yellow prussiate of soda.

antioxidants: the odor, taste, and color of most foods gradually deteriorate during storage. Most of the changes are caused by reaction of oxygen in the air with fats, carbohydrates, flavorings, and colorings. This reaction is known as oxidation.

The food constituents that deteriorate most rapidly are the unsaturated fats and oils. When these compounds react with oxygen they generate new substances that have offensive—rancid —tastes and odors.

Other substances that are unstable in the presence of oxygen include coloring matter, such as the beta carotene in carrots and margarine, and the carbohydrates in fruits and vegetables. Carotene turns gray upon oxidation, whereas carbohydrate, such as that which is exposed when an apple or potato is peeled, turns brown.

Several means have been devised to prevent or retard oxidation and thereby extend the storage life of foods. One method is to prevent food from coming in contact with air. This can be done by using vacuum packs or by packing food in the presence of nitrogen, an inert gas. A more convenient way of eliminating food-air contact is to add to the food a chemical, such as ascorbic or isoascorbic acids or stannous chloride, that combines with oxygen and depletes the oxygen supply in a container.

The most efficient inhibitors of oxidation are the familiar compounds BHA, BHT, and propyl gallate.[1] These "primary antioxidants" can interrupt the chain reaction oxidations, which mark the deterioration of fats.[2] Fats going rancid might be compared to a row of falling dominoes. The more dominoes that fall, the worse the smell. BHA, BHT, and propyl gallate prevent oxidation much like a finger holding back a falling domino would prevent a whole row of dominoes from toppling. The tocopherols (vitamin E) and gum guaiac are natural antioxidants which function in the same way.

Another method of preventing oxidative changes in food followed from the discovery that minute amounts of metals (copper, zinc, iron, etc.) accelerate oxidation. Calcium disodium EDTA, citric acid, sodium pyrophosphate, and other chemicals trap these metal ions and curb oxidation much as a vacuum cleaner traps dirt or filters remove impurities from fluids. These chemicals are called sequestrants or chelating agents. Chemicals of this type occur naturally and help retard oxidation in living plants and animals.

The food industry, fearful that the word antioxidant has taken on a pejorative meaning, has coined several euphemisms. "Freshness preserver" and "oxygen scavenger" are two examples.

The long shelf life of foods that contain antioxidants gives

---

[1] By the time this book is published, the food industry may be using yet another synthetic antioxidant: tertiary butylhydroquinone, or TBHQ. Eastman Chemicals, the purveyor of TBHQ, refused to send the author a summary of the toxicity tests they claim to have done, so the safety of this chemical cannot be judged.

[2] Oxygen reacts with fats to produce peroxides, which generate free radicals; the very reactive free radicals are removed by primary antioxidants.

bacteria and mold more opportunity to grow. For this reason antimicrobial preservatives are frequently used in conjunction with antioxidants.

Antioxidants are frequently added unnecessarily to foods (see the individual sections on butylated hydroxyanisole and butylated hydroxytoluene, vegetable oil).

baking powder: a chemical leavening system that substitutes for yeast. It consists of appropriate proportions of baking soda and one or more acids. Most household baking powders are of the "double-acting type," which contain two acids, sodium aluminum phosphate $[Al_2(SO_4)_3 \cdot Na_2SO_4]$ and monocalcium phosphate $[CaH_4(PO_4)_2]$. It is called double-acting because monocalcium phosphate forms gas cells during the preparation of the dough, while the other acid does not release gas until the dough is hot. The other important kind of baking powder, said to be the more efficient, contains only baking soda and monocalcium phosphate. Starch is added to baking powder to absorb moisture and prevent the soda and acid from entering into a chemical reaction on the shelf. Flour to which baking powder has been added is called "self-rising flour."

baking soda: sodium bicarbonate $(NaHCO_3)$. Carbon dioxide gas is liberated when baking soda and an acid are mixed in water. See baking powder.

bleaching agent: a chemical that makes flour white by oxidizing the colored matter (xanthophylls). See acetone peroxide, benzoyl peroxide, chlorine, chlorine dioxide. These agents destroy what little vitamin E (tocopherol) remains in white flour.

buffer: a chemical that maintains a solution or food at the desired acidity.

carcinogen: a substance that causes cancer.

cariogenic: promoting tooth decay.

chelator (derived from the Greek word meaning claw): a chemical that chelates, or traps, trace amounts of metal atoms that would otherwise cause food to discolor or go rancid. See EDTA, citric acid. Same as sequestrant.

co-carcinogen: a substance that by itself does not cause cancer but makes an animal more sensitive to certain carcinogens. The oil of croton seeds is the most commonly used laboratory co-carcinogen.

Delaney Clause: the sections of the 1958 Food Additive and the 1960 Color Additive amendments to the Food, Drug and Cosmetic Act that specifically prohibit in food any chemical that causes cancer when fed *at any level* to animals or man.

The clause is named after its chief congressional proponent, Representative James J. Delaney of New York.

detoxifying enzymes: enzymes in the liver that convert poisons into harmless chemicals, which are then excreted.

diastatic activity: ability to degrade starch to glucose and maltose; see amylases.

dough conditioner: a substance used by bakers to make dough drier (less sticky), more extensible, and easier to machine. See discussion of calcium stearoyl lactylate; see also aging agents.

emulsifiers: most liquids are either water-like or oil-like, and, as everyone knows, the two do not ordinarily mix. Both Nature and chemists, however, have developed chemicals that have both oil-like and water-like properties and that enable water and oil to mix. The mixing aids are known as emulsifiers (because they create emulsions) or surfactants (because they decrease surface tensions). They are close relatives of soaps and detergents.

Emulsifiers create emulsions, not true solutions. In other words, they stabilize and decrease the size of fat globules in water, or water globules in oil, to such an extent that the globules have little tendency to coalesce and form distinct layers. The way an emulsion would look under a microscope is shown in the figure.

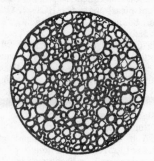

An oil—water emulsion

Emulsifiers are naturally present both in foods and in the body. The major difference between mayonnaise and oil-and-vinegar dressing, for instance, is the egg in the mayonnaise. The egg protein adds body, and the egg lecithin emulsifies the vegetable oil and vinegar so that they do not separate into two layers. Bile salts, which are produced by the liver and secreted into the small intestine, emulsify fats so that they can be absorbed more readily by the body.

The most frequently used emulsifiers in foods are mono- and diglycerides, which are derived from fat or oil, and the synthetic polysorbates and sorbitan monostearate. These and other agents keep bread from going stale by preventing starch molecules from crystallizing, help powdered nondairy creamers dissolve in coffee, prevent oil from separating out of peanut butter, make cakes light and fluffy, enable oil-like colorings and flavorings to mix well with foods, serve as whipping aids in toppings, keep fat from separating out in processed meat products and pet foods, and keep butterfat in solution during the initial freezing of ice cream. Most convenience and manufactured foods depend in one way or another on added emulsifiers.

Thickening agents (stabilizers) are sometimes called emulsifiers because they hinder the movement of oil or water globules and prevent them from forming separate layers.

Emulsifiers are rarely used to adulterate food, but their use in peanut butter may be an exception. Up to 10 percent of the peanuts may be replaced by emulsifiers to prevent peanut oil from separating out. "Ninety percent peanut butter" is considered adulterated by some consumers.

"enriched": products such as bread and spaghetti to which nutrients have been added (after naturally occurring nutrients have been removed) are labeled "enriched." Enrichment does not replace all the lost nutrients.

enzymes: specialized protein molecules which facilitate the interconversion, modification, degradation, or synthesis of molecules in living organisms. Many enzyme molecules consist of protein molecules combined with a vitamin molecule or a metal atom.

epidemiology: the study of the incidence of a disease in a population. Usually, the incidence is correlated with another variable, such as body weight, geographic location, coffee consumption, etc.

essential amino acids: amino acids which the body cannot synthesize from other molecules and which must be obtained from food. Isoleucine, leucine, lysine, methionine, phenylalanine, threonine, tryptophan, valine are essential amino acids for man.

FAO/WHO Expert Committee on Food Additives: an international committee, sponsored by the Food and Agriculture Organization and the World Health Organization of the United Nations, made up of government, industry, and university scientists. The committee reviews the safety of important food additives and then publishes evaluations and recommendations. The committee's work is of particular value to small nations that cannot afford to maintain extensive food safety agencies.

FDA: Food and Drug Administration, a division of the U. S. Department of Health, Education and Welfare. The FDA was established in 1906 to protect the public from unsafe foods, drugs, and cosmetics. In recent years the FDA has also been empowered to regulate toys and hazardous substances, such as laundry detergents.

fermentation: breakdown of starch, sugar, or fats, usually accompanied by the production of acids and other small molecules, by the action of bacteria, yeast, etc. The production of yogurt, cheese, and beer involve fermentation. See the discussion of citric acid for a description of how fermentation can be used in the manufacture of chemicals.

firming agent: manufacturers add calcium salts (calcium chloride, calcium citrate, calcium lactate, etc.) to canned tomatoes, potatoes, lima beans, and peppers to prevent them from becoming unacceptably soft. The calcium acts like a cement by reacting with the pectin in the cell walls of fruits and vegetables.

flavor enhancer: a substance that has little or no flavor of its own but brings out the natural flavor of foods. The mechanism by which flavor enhancers operate is not known. They may sensitize taste buds, stimulate the secretion of saliva (which breaks down food and lets the flavor out), etc. See monosodium glutamate (MSG), disodium guanylate and disodium inosinate, maltol and ethyl maltol.

fortified: manufacturers sometimes add vitamins and minerals to products that are low in nutrients, such as breakfast cereals and imitation fruit drinks, and label the products "fortified."

Because synthetic vitamins and minerals are very inexpensive, food fortification has become a highly effective, low-cost advertising gimmick. Snack foods and breakfast cereals, which many manufacturers fortify, contain large amounts of sugar and are basically bad foods. Hoffman-La Roche, the nation's major vitamin manufacturer, told potential customers that "Recent research indicates that more and more mothers everywhere are aware of the need for vitamin and mineral fortification. And are even willing to pay a little extra for it . . . we can show you why it pays to fortify" (*Food Product Development,* August/September 1971).

germs: the general name for viruses, bacteria, molds, protozoa, and other disease-causing microscopic organisms.

glycerolysis: degradation of triglycerides (fats and oils) to di- and monoglycerides.

gram: a unit of weight, approximately 1/28 of an ounce.

GRAS—Generally Recognized As Safe: a category of food additives established by the FDA. The GRAS (pronounced grass) category was created to exempt unquestionably safe chemicals, such as starch, salt, sugar, baking soda, protein, etc., from the costly toxicological tests required by the 1958 food additive law. Inevitably, though, chemicals that were not unquestionably safe joined (and are still on) the list.

According to the law, chemicals that were used in food prior to January 1, 1958, could be deemed GRAS without scientific testing. Chemicals introduced to the food supply after that date may not be declared GRAS unless scientific tests indicate safety. The catch is that the company that markets the additive can decide what constitutes adequate scientific testing.

GRAS chemicals differ from other food additives in three main respects:

1. GRAS chemicals are *assumed* to be safe because they have been used in food for many years; most other additives are thought to be safe because they have undergone a certain amount of scientific testing.

2. There is no limit on the levels at which most GRAS substances may be added to food; the FDA places limits on most other food additives, both as to concentration and as to the foods in which they may be used.

3. A manufacturer can declare a chemical GRAS and use it in food without consulting the FDA (the FDA can challenge a firm if it disagrees with the firm's judgment); other food additives must be tested—and the tests must be approved by the FDA—*before* they may be used in food.

Most GRAS chemicals have been used in foods for many years and are undoubtedly safe. But being on the GRAS list certainly does not guarantee safety, because many GRAS chemicals have been inadequately tested. Cyclamate was on the list in 1969 when it was found to be carcinogenic; brominated vegetable oil (BVO) was on the list in 1970 when it was found to be toxic.

The FDA announced in 1971 that it would re-examine the toxicity of all GRAS compounds and conduct tests on many of them. It is a safe bet that several will be removed from the list. Appendix 2 contains a partial list of GRAS compounds.

hemoglobin: the oxygen-carrying protein molecule in red blood cells. The red color of blood reflects the iron content of hemoglobin.

humectant: a chemical that is incorporated into a food (marshmallows, shredded coconut, candies, etc.) to maintain the desired level of moisture. See glycerol, propylene glycol, sorbitol.

hydrolysis: the splitting of a molecule (protein, carbohydrate, etc.) by reaction with water.

in vitro experiments: experiments performed with components of living organisms (tissue slices, cultured cells, cell sap, stomach juices, etc.).

in vivo experiments: experiments done on intact, living animals.

laxative: a substance that helps move the bowels. Mineral oil and similar substances lubricate the large intestine, thereby expediting movement of stools. Karaya and carboxymethylcellulose are bulk-type laxatives. They are not absorbed in the intestine and become part of the feces. These chemicals absorb a large amount of water, expand in size, and stimulate the intestine to move the stools.

LD_{50}—Lethal Dose 50 percent: the amount of a chemical that kills 50 percent of a group of animals. The lower the LD_{50}, the more poisonous the chemical. LD_{50} is a commonly used index of a chemical's toxicity, but it says nothing about carcinogenicity, mutagenicity, cumulative effects, etc. The LD_{50} for a given chemical may vary considerably from species to species. When discussing LD_{50}s the route of administration of the chemical must be indicated (oral, subcutaneous, etc.). LD_{50}s are usually expressed in units of mg/kg (i.e. milligrams of the chemical per kilogram of the test animal's body weight).

maturing agent: see aging agent.

metabolism: the degradation, synthesis, and interconversion of molecules by living organisms.

mg/kg: dosages of a chemical used in an experiment are usually expressed in units of milligrams (mg) of the chemical per kilogram (kg) of body weight of a laboratory animal or person. One milligram is approximately 1/28,000 of an ounce; one kilogram is 2.2 pounds.

minerals: the body needs inorganic mineral nutrients such as iron, zinc, cobalt, calcium, and copper. Metal atoms in living organisms are often combined with proteins and other molecules: hemoglobin contains iron, insulin contains zinc, vitamin B-12 contains cobalt.

mutation: an inheritable change in an organism's genetic material. The effects of mutations range from undetectably small changes in a cell, to hemophilia, to death. Mutations may be caused by radiation (X rays, cosmic rays, fallout) or certain chemicals. Agents that cause mutations are called mutagens.

myoglobin: hemoglobin-like molecules that store oxygen in muscle tissue.

nucleic acid: the molecules that contain, in chemically encoded form, the genetic information of all organisms. DNA (deoxyribonucleic acid) and RNA (ribonucleic acid) are the two kinds of nucleic acids.

pasteurization: the heat treatment of foods to kill contaminating microorganisms.

ppm: parts per million. A convenient unit used to denote how much of a chemical is present in a food or another chemical. Thus, a food may contain 200 parts (grams, ounces, pounds) of a preservative per million parts (grams, ounces, pounds) of food.

precipitate: when a chemical comes out of solution due to chemical or physical forces, it is said to precipitate. The curdling of milk (precipitation of casein) that occurs when it is treated with acid (mix milk with vinegar or grapefruit juice) is a familiar example of precipitation.

preservative: a chemical used to increase the safety or shelf life of a food. Antimicrobial preservatives inhibit the growth of microorganisms (see calcium propionate, parabens, sodium benzoate, sodium nitrate and nitrite, sulfur dioxide). Antioxidant preservatives prevent rancidity and discoloration (see butylated hydroxyanisole and butylated hydroxytoluene, citric acid, EDTA, propyl gallate, stannous chloride).

protein: muscle, hair, nail, cartilage, and most other parts of living organisms contain protein molecules. Enzymes are proteins. Protein molecules are composed of amino acids linked end to end in a specific order.

rancidity: the offensive odor that develops as food spoils; it is caused primarily by oxygen reacting with unsaturated fats and oils. See antioxidant.

rope: a condition in bread characterized by gelatinous threads that form in the center of a loaf. Rope is caused by certain spore-forming bacteria (*Bacillus mesentericus, B. subtilis*) that may contaminate dough and survive the baking process. As the bacteria multiply they digest the bread.

sequestrant: a chemical that sequesters or traps trace amounts of metal atoms that would otherwise cause food to discolor or go rancid; same as "chelator." See EDTA, citric acid.

stabilizer: see thickening agent.

subcutaneous: under the skin.

syndrome: the outward signs or symptoms of a disease; e.g. sneezing for hay fever; self-mutilation for Lesch-Nyhan disease; high fever, skin eruption, nasal catarrh for measles.

teratogen: a chemical that can cause birth defects (terata). *Teratos* is the Greek word for monster.

thickening agent (stabilizer):[3] manufacturers use thickening agents to "improve" the texture and consistency of ice cream, pudding, soft drinks, salad dressing, yogurt, soup, baby food and formula, and other foods. These chemicals control the formation of ice crystals in ice cream and other frozen foods. The thickness they create in salad dressing prevents the oil and vinegar from separating out into two layers. These additives are used to stabilize factory-made foods, that is, to keep the complex mixture of oils, acids, colors, salts and nutrients dissolved and at the proper consistency and texture.

Many companies substitute thickening agents for "tender loving care" and time-consuming production methods. Examples taken from the dairy industry illustrate this point. Cheaper

[3] See table following.

THICKENING AGENTS

Substance	Source	Composition (major species of carbohydrate sub-units)
agar	seaweed	D- and L-galactose (contains sulfate esters)
alginate	seaweed	D-mannuronic acid; L-guluronic acid
arabic (acacia) gum	tree exudate	L-arabinose; L-rhamnose; D-galactose; D-glucuronic acid
arabinogalactan	tree exudate	D-galactose; L-arabinose
carrageenan	seaweed	D-galactose; 3, 6-anhydro-D-galactose (contains sulfate esters)
cellulose, microcrystalline	plants	D-glucose
cellulose derivatives:	cellulose	D-glucose containing one or more different side chains
carboxymethylcellulose		
ethyl cellulose		
hydroxypropyl cellulose		
hydroxypropyl methyl cell		
methyl cellulose		
methyl ethyl cellulose		
furcelleran	seaweed	D-galactose; 3, 6-anhydro-D-galactose (contains sulfate esters)
ghatti gum	tree exudate	L-arabinose; D-xylose, galactose, mannose, glucuronic acid
guar gum	plant seed	D-mannose; D-galactose
karaya (sterculia) gum	tree exudate	D-galactose; L-rhamnose; D-galacturonic acid
locust bean (carob seed) gum	plant seed	D-mannose; D-galactose
pectin	fruits, berries	D-galacturonic acid (contains methyl esters)
propylene glycol alginate	seaweed derivative	D-mannuronic acid; L-guluronic acid; propylene glycol groups
starch	corn, potato, etc.	D-glucose
starch, modified	starch derivatives	oxidized, hydrolyzed, bleached, cross-linked, derivatized starch molecules
tragacanth gum	tree exudate	D-galactose; D-xylose; D-glucuronic acid
xanthan gum	microorganism	D-glucose; D-mannose; D-glucuronic acid

brands of ice cream and yogurt contain gelatin, carrageenan and other chemicals that prolong the shelf life and affect the consistency. A handful of quality-conscious manufacturers, on the other hand, use more demanding and costly production techniques instead of additives. Dannon and Colombo yogurts and Breyer ice cream, to cite just three examples, contain no artificial thickening agents (or other additives). These products prove that artificial thickeners are not vital ingredients in commercial yogurt and ice cream. Consumers who are willing to pay the higher price receive products which taste better and which carry minimum health risks.

Most thickening agents are natural carbohydrates (agar, carrageenan, pectin, starch, etc.) or chemically modified carbohydrates (cellulose gum, modified starch, etc.). They work by absorbing part of the water that is present in a food, thereby making the food thicker. The Table on page 248 lists the source and composition of the major thickening agents.

toxic: poisonous. Mutations, cancer, blindness, liver damage, etc., are typical of toxic effects that can be caused by chemicals.

toxicity tests: (a) acute toxicity tests reveal the effects of mammoth single doses of a chemical;

(b) short-term toxicity tests reveal the effects of 30–180-day exposure to a substance, with special attention usually given to liver and kidney function and blood composition;

(c) long-term toxicity tests last the lifetime of an animal (two years in rats or mice and seven years in dogs). These experiments are the only kind that can reveal whether a chemical causes cancer or chronic effects.

(d) other special tests are designed to detect interference with reproduction and causation of birth defects or mutations.

vitamins: chemicals that the body needs but cannot make and therefore must obtain from foods (or vitamin pills). Not all animals require the same set of vitamins. For instance, the only animal species that cannot make vitamin C (ascorbic acid) are primates (including humans), the guinea pig, one bird species, and one species of bat.

BIBLIOGRAPHY

References

C&EN: "Food Additives," Sanders, H. J., *Chem. Eng. News 44* October 10 and October 17, 1966.

CRC: Furia, T. E., ed., *Handbook of Food Additives,* Chemical Rubber Company, Cleveland, 1968.

CUFP: National Academy of Sciences–National Research Council, "Chemicals Used in Food Processing," Pub. No. 1274 (1965).

ECT: *Encyclopedia of Chemical Technology,* Interscience Publishers, New York, 2d edition, 1963–.

FAO(17): "Procedures for the Testing of Intentional Food Additives to Establish Their Safety for Use," FAO Nutrition Meetings Report Series No. 17, WHO Tech. Rep. Series 144 (1958).

FAO(29): "Evaluation of the Carcinogenic Hazards of Food Additives," FAO Nutrition Meetings Report Series No. 29, WHO Tech. Rep. Series 220 (1961).

FAO(31): "Evaluation of the Toxicity of a Number of Antimicrobials and Antioxidants," FAO Nutrition Meetings Report Series No. 31, WHO Tech. Rep. 228 (1962).

FAO(35): "Specifications for the Identity and Purity of Food Additives and Their Toxicological Evaluation: Emulsifiers, Stabilizers, Bleaching and Maturing Agents," FAO Nutrition Meetings Report Series No. 35, WHO Tech. Rep. Series 281 (1964).

FAO(38A): "Specifications for Identity and Purity and Toxicological Evaluation of Some Antimicrobials and Antioxidants," FAO Nutrition Meetings Report Series No. 38A (1965).

FAO(38B): "Specifications for Identity and Purity and Toxicological Evaluation of Food Colors," FAO Nutrition Meetings Report Series No. 38B (1966).

FAO(40): "Specifications for the Identity and Purity of Food Additives and Their Toxicological Evaluation: Some Antimicrobials, Antioxidants, Emulsifiers, Stabilizers, Flour-Treatment Agents, Acids and Bases," FAO Nutrition Meetings Report Series No. 40, WHO Tech. Rep. Series 339 (1966).

FAO(40A): "Toxicological Evaluation of Some Antimicrobials, Antioxidants, Emulsifiers, Stabilizers, Flour-Treatment Agents, Acids and Bases," FAO Nutrition Meetings Report Series No. 40 A, B, C (1967).

FAO(43): "Specifications for the Identity and Purity of Food Additives and Their Toxicological Evaluation: Some Emulsifiers and Stabilizers and Certain Other Substances," FAO Nutrition Meetings Report Series No. 43, WHO Tech. Rep. Series 373 (1967).

FAO(44A): "Toxicological Evaluation of Some Flavoring Substances and Non-nutritive Sweetening Agents," FAO Nutrition Meetings Report Series No. 44A (1967).

FAO(46): "Specifications for the Identity and Purity of Food Additives and Their Toxicological Evaluation: Some Food Colors, Emulsifiers, Stabilizers, Anticaking Agents, and Certain Other Substances," FAO Nutrition Meetings Report Series No. 46, WHO Tech. Rep. Series 445 (1970).

FAO(46A): "Toxicological Evaluation of Some Food Colors, Emulsifiers, Stabilizers, Anti-Caking Agents and Certain Other Substances," FAO Nutrition Meetings Report Series 46A (1970).

Related Readings

Ayres, J. C., Kraft, A. A., Snyder, H. E., Walker, H. W. (ed.), *Chemical and Biological Hazards in Food,* Iowa State University Press, Ames, Iowa, 1962.

Carson, Rachel, *Silent Spring,* Houghton Mifflin, Boston, 1962.

Davis, Adelle, *Let's Cook It Right,* New American Library, New York, 1970.

Goodwin, R., *Chemical Additives in Food,* Churchill, London, 1967.

Graham, F., *Since Silent Spring,* Houghton Mifflin, Boston, 1970.

Lappé, M., *Diet for a Small Planet,* Ballantine, New York, 1972.

Margolius, S., *The Great American Food Hoax,* Walker & Co., New York, 1971.

National Academy of Sciences, "Evaluating the Safety of Food Chemicals," Publication No. 1859, Washington, 1970.

Nutrition Foundation, *Present Knowledge in Nutrition,* 3d ed., New York, 1967.

Turner, J., *The Chemical Feast,* Grossman, New York, 1970.

U. S. Dept. of Health, Education and Welfare, "Report of the Secretary's Commission on Pesticides and Their Relationship to Environmental Health" (the Mrak Report), Washington, 1969.

Wellford, H., *Sowing the Wind,* Grossman, New York, 1972.

INDEX